Measuring Discharge with Acoustic Doppler Current Profilers from a Moving Boat

By David S. Mueller and Chad R. Wagner

Chapter 22 of Book 3, Section A

Techniques and Methods 3–A22

U.S. Department of the Interior
U.S. Geological Survey

U.S. Department of the Interior
DIRK KEMPTHORNE, Secretary

U.S. Geological Survey
Mark D. Myers, Director

U.S. Geological Survey, Reston, Virginia: 2009

For product and ordering information:
World Wide Web: http://www.usgs.gov/pubprod
Telephone: 1-888-ASK-USGS

For more information on the USGS—the Federal source for science about the Earth, its natural and living resources, natural hazards, and the environment:
World Wide Web: http://www.usgs.gov
Telephone: 1-888-ASK-USGS

Suggested citation:
Mueller, D.S., and Wagner, C.R., 2009, Measuring discharge with acoustic Doppler current profilers from a moving boat: U.S. Geological Survey Techniques and Methods 3A–22, 72 p. (available online at http://pubs.water.usgs.gov/tm3a22).

Foreword

The mission of the U.S. Geological Survey (USGS) Water Resources Discipline is to provide the information and understanding needed for wise management of the Nation's water resources. Inherent in this mission is the responsibility of collecting data that accurately describe the physical, chemical, and biological attributes of water systems. These data are used for environmental and resource assessments by the USGS, other government agencies and scientific organizations, and the general public. Reliable and quality-assured data are essential to the credibility and impartiality of the water-resources appraisals carried out by the USGS.

The development and use of guidelines for Measuring Discharge with Acoustic Doppler Current Profilers from a Moving Boat are necessary to achieve consistency in the use of scientific methods and procedures, document the methods and procedures used, and maintain technical expertise in the process. USGS hydrographers and hydrologists can use this manual to ensure that the data collected are of the quality required to fulfill our mission.

Measuring Discharge with Acoustic Doppler Current Profilers from a Moving Boat contains the most current information and guidance regarding acoustic Doppler current profilers (ADCPs) used by the USGS at the time of publication. The development of new and improved ADCPs is ongoing, as are the research and practical field experience with existing and new ADCPs, which likely will lead to changes in the guidance on the application of ADCPs over time and revisions to this document. The user is encouraged to log onto the USGS Office of Surface Water Web site [http://hydroacoustics.usgs.gov] for the latest revisions to this document and technical memorandums that may be issued prior to revisions to ensure that the best techniques are communicated for use in collecting and processing ADCP discharge measurements.

Stephen Blanchard
Chief, Office of Surface Water

Contents

Figures

Tables

Conversion Factors

Inch/Pound to SI

Multiply	By	To obtain
Length		
foot (ft)	0.3048	meter (m)
mile (mi)	1.609	kilometer (km)
Flow rate		
foot per second (ft/s)	0.3048	meter per second (m/s)
cubic foot per second (ft³/s)	0.02832	cubic meter per second (m³/s)

SI to Inch/Pound

Multiply	By	To obtain
Length		
centimeter (cm)	0.3937	inch (in.)
millimeter (mm)	0.03937	inch (in.)
meter (m)	3.281	foot (ft)

Temperature may be converted as follows:

$$°F = (1.8 \times °C) + 32$$
$$°C = (°F - 32) / 1.8$$

Abbreviations and acronyms used in this report:

ABS	acoustic backscatter
ADCP	acoustic Doppler current profiler
ADP	acoustic Doppler profiler
BB	broadband
CD–ROM	compact disc–read-only memory
CMG	course made good
DBT	depth below transducer
DC	direct current
DGPS	differentially corrected global positioning system
DMG	distance made good
DOP	dilution of precision
EMF	electromagnetic field
FAA	Federal Aviation Administration
GPS	global positioning system
HDOP	horizontal dilution of precision
Hz	hertz
kHz	kilohertz
LAN	local area network
MHz	megahertz
NMEA	National Marine Electronics Association
OSW	Office of Surface Water
PCMCIA	Personal Computer Memory Card International Association
PDA	portable digital assistant
ppt	parts per thousand
RTK	real-time kinematic
SMBA	Stationary Moving-Bed Analysis
TRDI	Teledyne RD Instruments
USB	Universal Serial Bus
USGS	U.S. Geological Survey
WAAS	Wide Area Augmentation System
WM	water mode
WM12sp	water mode 12 in StreamPro
WP	water ping
WV340	ambiguity velocity of 340 millimeters per second

Measuring Discharge with Acoustic Doppler Current Profilers from a Moving Boat

By David S. Mueller and Chad R. Wagner

Abstract

The use of acoustic Doppler current profilers (ADCPs) from a moving boat is now a commonly used method for measuring streamflow. The technology and methods for making ADCP-based discharge measurements are different from the technology and methods used to make traditional discharge measurements with mechanical meters. Although the ADCP is a valuable tool for measuring streamflow, it is only accurate when used with appropriate techniques. This report presents guidance on the use of ADCPs for measuring streamflow; this guidance is based on the experience of U.S. Geological Survey employees and published reports, papers, and memorandums of the U.S. Geological Survey. The guidance is presented in a logical progression, from predeployment planning, to field-data collection, and finally to post-processing of the collected data. Acoustic Doppler technology and the instruments currently (2008) available also are discussed to highlight the advantages and limitations of the technology. More in-depth, technical explanations of how an ADCP measures streamflow and what to do when measuring in moving-bed conditions are presented in the appendixes. ADCP users need to know the proper procedures for measuring discharge from a moving boat and why those procedures are required, so that when the user encounters unusual field conditions, the procedures can be adapted without sacrificing the accuracy of the streamflow-measurement data.

Introduction

The acoustic Doppler current profiler (ADCP) has evolved during the last 25 years from an experimental instrument capable of measuring velocity and computing discharge in deep water (greater than 11 feet (ft)) to an instrument that is commonly used to measure water velocity and discharge in streams as shallow as 1.0 ft deep (Christensen and Herrick, 1982; Simpson and Oltmann, 1993; Oberg and Mueller, 2007b). The development of the ADCP has provided hydrographers and engineers with a tool that can substantially reduce the time for making discharge measurements and can measure water velocities at a spatial and temporal scale

that was previously unattainable. These instruments are used regularly to measure riverine and estuarine water discharge, to collect data for hydrodynamic model calibration and verification, to assess aquatic habitat, and to study sediment transport processes. Although the use of the ADCP has become common, proper instrument configuration and data-collection and post-processing procedures are required to collect accurate and reliable data.

Purpose and Scope

The purpose of this report is to present the procedures that should be followed when using an ADCP from a moving boat to make surface-water discharge measurements. The procedures for predeployment preparation, field data collection, and processing of collected data are discussed. A detailed description of how an ADCP measures velocity and computes discharge and additional details on selected topics are presented in appendixes.

Applications

The measurement of unsteady, bidirectional, and other flows with nonlogarithmic velocity distributions has been a problem faced by hydrologists for many years. Dynamic discharge conditions impose an unreasonable time constraint on conventional current-meter discharge-measurement methods, which typically take at least 1 hour to complete. Tidally affected discharge can change more than 100 percent during a 10-minute period. In addition, bidirectional flows caused by density currents are common in tidally affected areas and have been increasingly observed in freshwater environments where a significant temperature gradient causes a density current (García and others, 2007). Nearly all discharge measurements made using point-velocity meters have an assumed standard logarithmic distribution of the horizontal velocity in the water column; however, wind-driven currents and very rough bottoms in shallow water may produce nonstandard profiles. The introduction of the ADCP into the coastal and riverine environments has enabled the development of a discharge-measurement system capable of more efficiently and more accurately measuring flow in unsteady, bidirectional, and nonstandard conditions. In most cases, an ADCP discharge-measurement system is faster than conventional discharge-measurement systems and has comparable or better accuracy because ADCPs measure a much larger portion of the water column than conventional discharge-measurement systems. More efficient discharge measurements improve safety by reducing the amount of time a hydrographer is on a bridge, on a boat, or in the water. The reduction in measurement time realized by using an ADCP is especially beneficial when trying to develop an index velocity rating (Morlock and others, 2002; Ruhl and Simpson, 2005) at sites with rapidly changing flow conditions. An ADCP can define the rating in the transitional range of flow that was otherwise indefinable with conventional discharge methods. In addition to measuring streamflow, ADCPs are used in a variety of other applications including:

- measurement of velocity fields for calibration of numerical models, hydraulic studies (for example, safety zones near dams), and habitat assessments;

- in situ deployments for current measurements and for aiding navigation;

- hydrographic surveys to measure channel bathymetry for use in hydrodynamic and habitat modeling applications; and

- estimation of sediment concentration from acoustic backscatter (ABS).

The application of acoustic technology in rivers and lakes has provided data that prior to the mid-1990s would have been unavailable or extremely expensive and impractical to collect.

Discussion of Instruments

The ADCP uses sound to measure water velocity. The sound transmitted by the ADCP is in the ultrasonic range (above the range heard by the human ear). The lowest frequency used by commercial ADCPs is around 30 kilohertz (kHz), and the common range for riverine measurements is between 300 and 3,000 kHz. The ADCP measures water velocity using a principle of physics discovered by Christian Johann Doppler (1842). Doppler's principle relates the change in frequency of a source to the relative velocities of the source and the observer. An ADCP applies the Doppler principle by reflecting an acoustic signal off small particles of sediment and other material (collectively referred to as scatterers) that are present in water. The velocity measured by the Doppler principle is parallel to the direction of the transducer emitting the signal and receiving the backscattered acoustic energy. Typical boat-mounted ADCPs have three or four beams pointing between 20 and 30 degrees from the vertical. Three beams are required to obtain a three-dimensional velocity

measurement. If a fourth beam is present, an additional error velocity can be measured (Appendix A).

In a boat-mounted system, the transducers are deployed beneath the water surface and aimed downward (fig. 1). Measurement of water velocity from a moving boat will yield the velocity of the water relative to the boat. ADCPs used in this manner account for the velocity of the boat by bottom tracking or through the use of a global positioning system (GPS). Bottom tracking determines the velocity of the boat by measuring the Doppler shift of acoustic signals reflected from the streambed; therefore, the water velocity relative to a fixed reference is computed by correcting the measured water velocity with the measured boat velocity.

Currently (2008) ADCPs can be classified into two groups based on the techniques used to configure and process the acoustic signal—narrowband and broadband. Narrowband is typically used in the hydroacoustic industry to describe a pulse-to-pulse incoherent ADCP; however, the narrowband ADCPs also can operate in a pulse-to-pulse coherent mode for short ranges. This means that in a narrowband ADCP, only one simple pulse is transmitted into the water, per beam per measurement (ping), and the resolution of Doppler shift takes place during the duration of the received pulse. This characteristic results in a system that is simple to configure and operate, but the velocity measurements made using the narrowband technology are noisy (have a relatively large random error). Narrowband systems compensate for the large random error by pinging fast (up to 20 hertz (Hz)) and averaging many pings together before reporting a velocity. Typical response from a narrowband system is a velocity-profile measurement every 5 seconds.

Broadband systems use a ping consisting of two or more synchronized acoustic pulses that are encoded with a pseudo-random code. The encoded pulse allows multiple velocity measurements to be made with a single ping, thus reducing the random noise associated in the measured velocity.

Figure 1.   Boat-mounted acoustic Doppler current profiler (ADCP).

Broadband systems are more difficult to configure because of the effect of the lag between the two pulses and because the processing of the complex pulse is slower than a narrowband system; however, the complex pulse results in a much lower random error, and the pulse pair allows configuration of the instrument to minimize random error for particular measurement conditions.

Predeployment Preparation

Prior to collecting data with an ADCP, it is important to establish standard procedures to ensure that the data collected will be stored in an efficient and consistent manner, the ADCP is in proper working order, and the ADCP is the appropriate equipment for making the measurement. Proper preparation will help avoid delays in the field, ensure complete and accurate data collection, and produce data that are documented and retrievable for future use. The required predeployment procedures include:

1. establishing a policy for handling and storing the data;

2. ensuring that ADCP hardware and software are working properly and are configured consistent with the policies established by the U.S. Geological Survey (USGS), Office of Surface Water (OSW);

3. identifying other equipment, such as a GPS and boats, that may be needed;

4. ensuring that the ADCP is capable of measuring the desired data for the expected field conditions; and

5. gathering and checking the ADCP and all ancillary equipment for use with the ADCP.

Data Management

The ADCP and associated software can produce a large number of files. It is important that these files are stored in a manner that allows users to easily identify the location, date, and type of data stored in the files. Because of the volatility of digital data, appropriate backup and archival procedures should also be implemented.

Naming Convention

Each office should establish and document a consistent naming convention for data files. Names of files should always be unique and should be descriptive of the data contained. Site number, site name, measurement number, project name, project number, and date are some of the descriptive terms that could be used in a filename. Typically, ADCP data-collection software will add a suffix to the user-defined name to identify the type of data file (configuration, raw data, ASCII data, etc.) and to ensure that each file has a unique name.

Data Storage and Archival

Each office collecting electronic discharge-measurement data needs to have a written policy about permanent file storage and archiving procedures. Procedures outlined are based on the assumption that an office has existing systems and procedures for performing routine backups and permanent archival for electronic information stored on servers. This policy should detail file and directory naming conventions, server directory structure, how soon data must be placed on the server after it is collected, and how, when, and where server data will be archived on stable archival media. USGS, Office of Surface Water Technical Memorandum 2005.08 (U.S. Geological Survey, 2005a) provides specific archival guidance. Paper measurement notes associated with an electronic discharge measurement should be filed and archived with other paper discharge-measurement notes in accordance with current office policies and procedures.

Each discharge measurement with electronic data files should have its own directory that contains all of the files collected or created as part of the measurement. These files include, but are not limited to, raw data, configuration information, moving-bed tests, instrument checks, calibration information, and discharge-measurement notes. The naming convention for the directories in the archival directory structure should include some combination of measurement number, measurement dates, water year, location, and(or) instrument types.

Instrument and Site Considerations

Any site-specific information, such as maximum water depths and velocities from previous measurements, can be used as a guide for configuring the ADCP for the measurement site. Notes about conditions and locations from previous ADCP discharge measurements should be reviewed prior to the field trip.

Limitations of ADCPs

The physics associated with sound generation from a transducer and then propagation, absorption, attenuation, and backscatter in the water column result in specific limitations and characteristics of ADCPs. Limitations that will be discussed in this report include the effect of sediment on backscattered acoustic energy and bottom tracking, and unmeasured areas of a profile associated with transducer draft and ringing and side-lobe interference. Additional limitations are imposed on ADCP measurements by the techniques used to configure and process the acoustic signal, which vary based on specific user configuration of the instrument.

Effect of Sediment

The quantity and characteristics of the particulate matter (such as sediment and aquatic life) in the water column can significantly affect the ability of the ADCP to make an accurate velocity measurement. Pure water is acoustically transparent because it has no suspended particulate matter to reflect acoustic energy. For a velocity measurement to be made, water must contain enough particulate matter for sufficient acoustic energy to be returned to the ADCP. Therefore, in very clear streams it is possible to have insufficient material in the water column to allow an ADCP to measure water velocity. High sediment loads, which are often present during high-flow conditions, can have the opposite effect. High sediment concentrations near the streambed can cause the ADCP to have trouble discriminating the streambed from the suspended-sediment concentration near the streambed, resulting in inaccurate water depth and(or) invalid boat velocity measurements (fig. 2A). In addition, high sediment concentrations in the water column can cause the acoustic signal to be attenuated before it can travel through the water column and back to the transducer, thus preventing the ADCP from making a measurement (fig. 2B). The sediment concentrations that trigger these limitations on ADCP operation have been observed but not quantified; these limitations depend on the sediment characteristics and on the water depth. In general, lower frequency acoustic instruments transmit more energy into the water and, therefore, are more capable of penetrating high sediment concentrations than higher frequency instruments.

During high flows, sediment transport near and along the streambed can cause a bias in the boat velocity determined from bottom tracking. Bottom tracking is used to determine the boat velocity and assumes that the streambed is stationary. Sediment transport near and along the streambed can cause a Doppler shift in the bottom-tracking ping and result in the boat-velocity measurement being biased in the upstream direction. This phenomenon is commonly referred to as a moving bed. If an ADCP is held stationary in a stream with a moving bed, a trace of the instrument motion based on bottom tracking shows the instrument moving upstream rather than being stationary. The result of a moving bed is that measured velocities and discharges will be biased low. Higher frequency instruments are more susceptible to moving-bed problems than are lower frequency instruments. Currently, there is no quantitative guidance for when a moving bed will be detected by an instrument, but tests to detect a moving bed are available and discussed later in this report. If a moving bed is detected, and the instrument is equipped with a compass, the use of GPS for measuring boat velocity is recommended. If the use of a GPS is not possible because of unfavorable site conditions, or if the instrument does not have a compass, other means to correct the discharge for the bias caused by the moving bed are available (Appendix B).

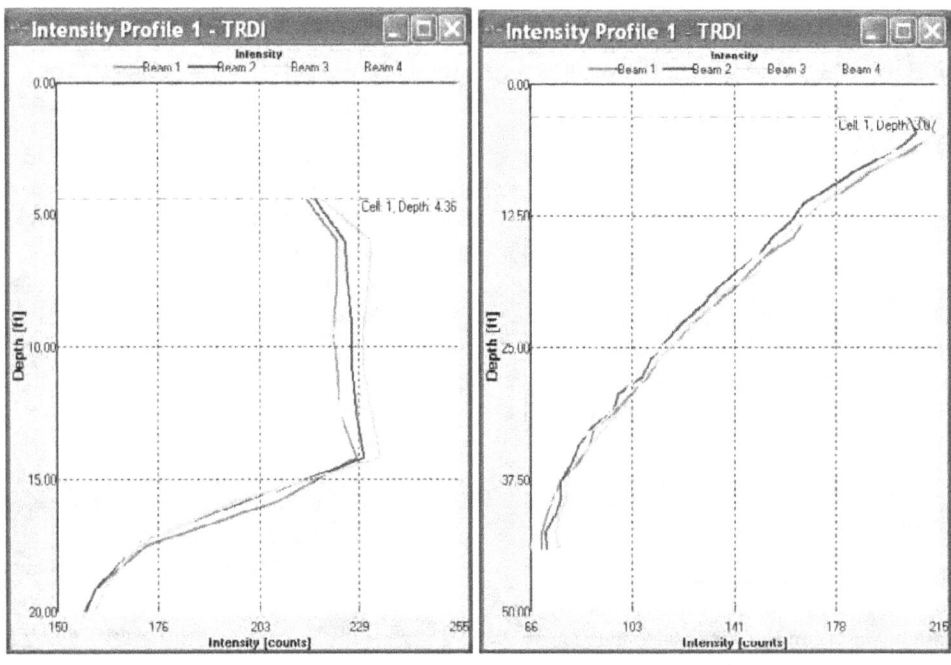

A) Excessive Backscatter   B) Excessive Attenuation

Figure 2.   Examples of (A) excessive backscatter and (B) attenuation due to sediment in the water as displayed in intensity-profile graphs from WinRiver II.

Unmeasured Areas in a Profile

ADCPs are called profilers because they provide measurements of velocity throughout the water column. The ADCP divides the water column into depth cells (also referred to by some software and references as bins) and reports a velocity for each depth cell; however, an ADCP cannot measure velocities at the water surface due to the draft of the instrument and the required blanking distance, nor can it measure near the bed due to side-lobe interference (fig. 3).

The length of the unmeasured area at the water surface is due to the draft of the instrument deployment, the effect of the transducer mechanics, and the flow disturbance around the instrument. The ADCP must be deployed below the water surface and, thus, cannot measure the water velocity above the transducers. The required instrument draft is controlled by the need to prevent the instrument from coming out of the water and to prevent entrained air from traveling under the instrument; therefore, the required instrument draft depends on the shape of the instrument mount, the boat, and the relative water velocity (water velocity past the instrument).

ADCPs use the same transducers to transmit and receive sound. When a transducer is energized to transmit sound, it vibrates to produce the sound waves. When the energy to the transducer is stopped, that transducer does not stop vibrating immediately; the vibrations dampen with time. The continued vibration of the transducer is called ringing and may be affected by the transducer housing and the ADCP mount. A good analogy of this effect is a large gong. The vibrations from a gong sometimes take several minutes to die out. The vibrations in a transducer die out much quicker than a gong, but sound travels some distance during the time it takes for the ringing to be reduced to a level where the transducer can accurately record backscattered acoustic signals. The distance that sound travels during the time it takes the ringing to be reduced is the minimum blanking distance. Depending on the frequency (typically lower frequency instruments have longer blanking distances) and the transducer housing, the blanking distance can vary from 0.16 to 3.3 ft. The flow disturbance caused by the instrument and its mount may also be a limiting factor of how close to the instrument an unbiased measurement of velocity can be made. Results from field data and numerical modeling suggest that for typical deployments, a blank of 0.82 ft (25 centimeters (cm)) for Teledyne RD Instruments (TRDI) Rio Grandes, 0.1 ft (3 cm) for TRDI StreamPros, and 0.66 ft (20 cm) for 3 megahertz (MHz) SonTek/YSI RiverSurveyors are acceptable; however, the deployment method and mount can influence the extent of the flow disturbance (Mueller and others, 2007).

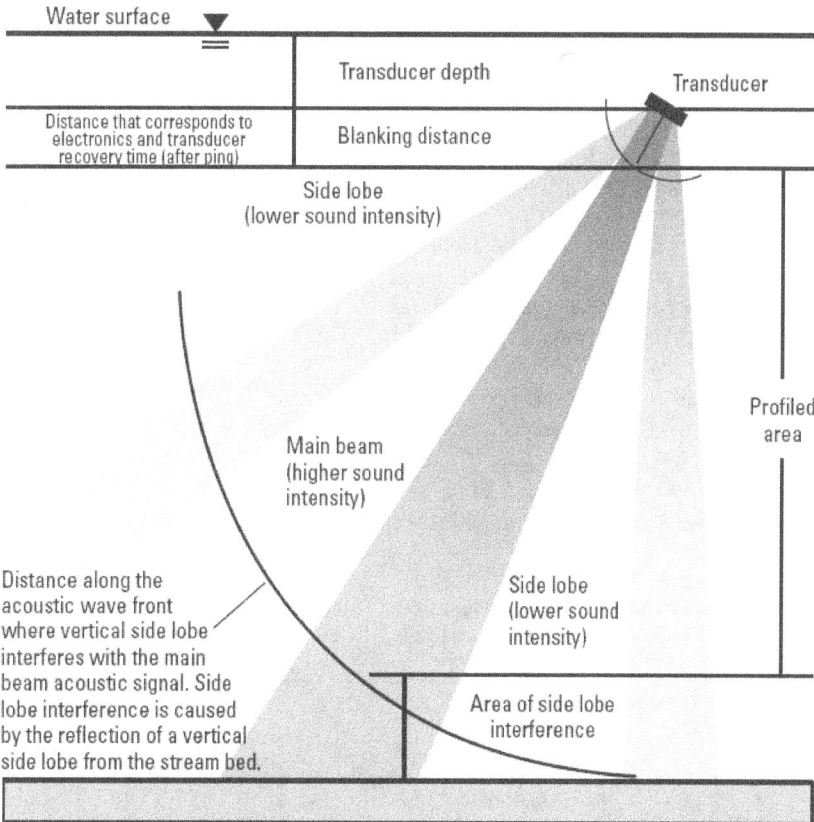

Figure 3.   Acoustic Doppler current profiler beam pattern and locations of unmeasured areas in each profile (from Simpson, 2002).

ADCPs cannot measure the water velocity near the streambed due to side-lobe interference (fig. 3). Most transducers that are developed using current (2008) technology emit parasitic side lobes off of the main acoustic beam. The acoustic energy in the side lobes is much less than in the main beam. The amount of acoustic energy backscattered from scatterers in the water column in the main beam is very small compared to the energy transmitted. The streambed reflects a much higher percentage of the acoustic energy than the scatterers in the water column. The magnitude of the energy in a side-lobe reflection from the streambed is sufficiently close to the energy reflected from scatterers in the main beam to cause potential errors in the measured Doppler shift. The water column affected by this side-lobe interference varies from 6 percent for a 20-degree system to 13 percent for a 30-degree system and can be computed as,

$$D_{SL} = D * (1 - \cos(\theta)), \qquad (1)$$

where

D_{SL} is the distance from the streambed affected by side-lobe interference;

D is the distance from the transducer to the streambed; and

θ is the angle of the transducers from the vertical.

The frequency and the techniques used to configure and process the acoustic signal are important in determining the maximum and minimum water depths that can be measured. Lower frequency ADCPs typically can measure deeper than higher frequency ADCPs but also require larger depth cells and a longer blanking distance. The operational mode of some ADCPs determines the location of the first and last valid depth cells and the acceptable size of the depth cells. The ADCP cannot measure the velocity in the upper and lower portions of the water column because of the draft, blanking distance, and side-lobe interference; therefore, the discharge in these areas must be estimated from data collected in the measured portion of the water column. For this reason, it is recommended that a

minimum of two depth cells be measured in the water column. The shallow-water limitation of an instrument is, therefore, the summation of the draft, blanking distance, location of the first depth cell, location of the last depth cell, the depth-cell size, and the range of the side-lobe interference.

Configuration and Characteristics

Site conditions, such as stream depth, water velocity, and bed material, ultimately dictate the instrument setup that will provide the most accurate discharge measurement. Currently, narrowband ADCPs do not have specific water or bottom modes that the user needs to select and configure. The primary setup for the narrowband instruments is setting the blanking distance and depth-cell size. The maximum profiling depth, maximum relative velocity, minimum recommended depth-cell size, and approximate random noise (velocity standard deviation) for SonTek/YSI RiverSurveyor ADCPs are presented in table 1.

Broadband ADCPs manufactured by TRDI offer multiple water and bottom modes. Although the multiple water and bottom modes make setup of the instrument more complicated, it allows the instrument to be optimized for the site conditions. Data-collection software for the Rio Grande ADCP from TRDI (Teledyne RD Instruments, 2003, 2007) has an automated configuration wizard that optimizes the instrument setup on the basis of the maximum expected velocity, boat speed, water depth, and bed material type.

Water modes offered in Rio Grande ADCPs allow the instrument to be optimized for the water velocity, depth, and bed material present at the time of the measurement. Each water mode has advantages and disadvantages associated with it. Water mode 1 is a robust multipurpose mode that can work in nearly all conditions, but the random noise associated with this mode limits the practical application in shallow, low-velocity situations. Water modes 5 and 11 are designed for low-velocity (less than 3.3 feet per second (ft/s)), shallow-water (less than 13 to 26 ft, depending on frequency) situations and have specific velocity and depth limitations. The advantages of water modes 5 and 11 are very low random

Table 1. Characteristics of SonTek/YSI RiverSurveyor acoustic Doppler current profilers.

[kHz, kilohertz; ft, foot; ft/s, foot per second; m, meter; data from SonTek, 2000]

Frequency (kHz)	Maximum profiling depth,[a] in ft	Maximum relative velocity,[b] in ft/s	Minimum recommended depth-cell size, in ft	1-second standard deviation, in ft/s
500	390	32	3.28 (1 m)	0.73
1,000	130	32	0.82 (0.25 m)	0.89
1,500	80	32	0.82 (0.25 m)	0.68
3,000	20	32	0.49 (0.15 m)	0.38

[a] The actual maximum depth that can be profiled depends on the water temperature and sediment in suspension.

[b] The maximum velocity measured by the acoustic Doppler current profile, which includes the boat and water speeds.

errors and small depth-cell sizes. Water mode 12 is a fast ping-rate mode that is similar to water mode 1, but uses a faster ping rate and an internal averaging technique of multiple pings per ensemble to reduce the random noise associated with normal mode 1 measurements. This reduction in noise by mode 12 allows smaller depth cells to be used or lower velocities to be measured with greater accuracy. The heading, pitch, and roll sensors are only measured at the beginning of the averaging interval, and bottom-track measurements do not occur during the averaging interval; therefore, random instrument movements caused by poor boat operation or turbulent water-surface conditions are unaccounted for in mode 12 and can cause significant errors if the averaging interval is too long. A maximum averaging interval of 1 second is recommended, and this interval may need to be further reduced in fast, turbulent conditions. The maximum profiling depth, the maximum relative velocity, recommended minimum depth-cell size, and random noise for the various water modes available in TRDI Rio Grande ADCPs are summarized in table 2. It is possible to collect valid data when the maximum relative velocity is greater than the values given in table 2; however, this is not necessarily predictable. These values should be used as a guideline to help users decide to use water mode 5 or 11. The high-resolution pulse coherent water modes 5 or 11 should be used wherever possible. It is important to also note that not every Rio Grande ADCP has water mode 12; it must be purchased separately and installed on the ADCP. Users should check their instruments to determine the available water modes by connecting to the ADCP with a terminal program (such as BBTalk, Teledyne RD Instruments, 2006) and issuing a "WM?" command. An in-depth discussion of the various water modes and their applicability to various site conditions can be found in Appendix C.

Currently (2008) the broadband ADCPs from TRDI have two bottom modes available. Bottom mode 5 is the general purpose and default water mode, but it does not work well in depths of less than approximately 2.6 ft below the transducer. Bottom mode 7 uses multiple lags to function in depths as shallow as 1.0 ft below the transducer and can function to the full maximum depth of the profiler. The bottom mode 7 multiple lag technique is slower, resulting in less data collected in a fixed time; therefore, bottom mode 7 typically is used only when bottom mode 5 fails to bottom track.

Two water modes are available for the TRDI StreamPro ADCP (table 3), which is designed for use in shallow water (less than 13 ft) with velocities less than 6.6 ft/s, measured with the standard integrated float (the instrument can physically measure up to 16 ft/s). In the default configuration, the StreamPro ADCP is limited to twenty 0.33-ft (10-cm) depth cells for a maximum profiling depth of 6.6 ft. A long-range mode is available for the StreamPro ADCP that increases the maximum depth-cell size to 0.66 ft (20 cm), extending the maximum water depth to 13 ft. The default mode is similar to water mode 12, which was explained previously. The Stream-Pro ADCP, however, does not actually use water mode 12 as implemented in the Rio Grande ADCP; rather, water mode 12 in the StreamPro ADCP (referred hereafter as WM12sp) is a modified multi-ping water mode 1, which pings fast. An

Table 2. Characteristics of Teledyne RD Instruments Rio Grande water-profiling modes for 1,200- and 600-kilohertz acoustic Dopppler current profilers.

[kHz, kilohertz; ft, foot; ft/s, foot per second; cm, centimeter; 600-kHz values are in brackets; data from Teledyne RD Instruments, 2008]

Water mode	Maximum depth,[a] in ft	Maximum relative velocity,[b] in ft/s	Minimum recommended depth-cell size, in ft	1-second standard deviation, in ft/s
1	65 [230]	32 [32]	0.82 [1.64] (25 [50] cm)	0.31 [0.31][c]
5[d]	13 [26][e]	2.3 [3.3][f]	0.16 [0.33] (5 [10] cm)	<0.03 [<0.03][c]
11[d]	13 [26][e]	2.3 [3.3][f]	0.16 [0.33] (5 [10] cm)	<0.03 [<0.03][c]
12	65 [230]	32 [32]	0.16 [0.33] (5 [10] cm)	0.59 [0.59][g]

[a] The actual maximum depth that can be profiled depends on the water temperature and sediment in suspension.

[b] The maximum velocity measured by the acoustic Doppler current profiler, which includes the boat and water speeds.

[c] Assumes a 2-hertz ping rate.

[d] Values are approximate.

[e] It is possible to profile deeper by decreasing the ambiguity velocity to 0.01 foot per second (WZ03), but this change reduces the maximum velocity. The WZ03 should be used with caution.

[f] The maximum velocity for modes 5 and 11 are highly dependent on depth and turbulence.

[g] Assumes 100 depth cells and an ambiguity velocity of 5.75 ft/s (WV175).

Table 3. Characteristics of Teledyne RD Instruments StreamPro acoustic Doppler current profiler water modes.

[ft, foot; ft/s, foot per second; cm, centimeter]

Water mode	Maximum depth, in ft	Maximum relative velocity, in ft/s	Minimum recommended bin size, in ft	1-second standard deviation, in ft/s
12sp	6.6 (13)[a]	16 (6.6)[b]	0.06	0.66[c]
13	3.3	0.82	0.06	0.006–0.066[d]

[a] Maximum depth of 13 ft requires purchase of extended range mode.

[b] Instrument should measure velocities of 16 ft/s, but the integrated float was designed for velocities less than 6.6 ft/s with a flat water surface.

[c] Standard deviation is for 0.07 ft (2 cm) depth-cell size. For 0.66 ft (20 cm) depth-cell size, the standard deviation is approximately 0.1 ft/s.

[d] Standard deviation depends on signal correlation, velocity, and depth.

ambiguity velocity of 11 ft/s (WV340) is used, making an ambiguity error nearly impossible for the StreamPro ADCP applications. The StreamPro ADCP has a second water mode (water mode 13 (WM13)) that has less random noise than WM12sp and can be used to measure water velocities less than about 0.82 ft/s in water less than 3.3 ft deep. WM13 is a long-lag pulse coherent mode. WM13 only becomes a selectable alternative for the user when the site conditions meet the criteria for maximum depth (less than 3.3 ft) and maximum velocity (less than 0.82 ft/s).

The bottom-tracking algorithm for the Streampro ADCP is different than the algorithms used for the Rio Grande ADCP. Each ensemble contains two bottom-track pings—one at the beginning of the ensemble and one at the end of the ensemble. The placement of the pings cannot be changed by the user.

Compass Considerations

Most ADCPs reference the water and boat velocity to magnetic north using an internal fluxgate compass. The effect of compass errors on measurements made with an ADCP is different for water-velocity and discharge data depending on the boat-velocity reference. When bottom tracking is used for the boat-velocity reference, a compass error will cause a rotational error in the measured water velocity, but the magnitude of the velocity is unaffected. The compass has no effect on measured discharge using bottom tracking as the boat-velocity reference; however, when an external boat-velocity reference such as GPS is used, the effect of the compass is substantial. Potential errors include errors in the compass reading caused by distortion of the Earth's magnetic field due to objects on the boat, displacement of the compass out of the horizontal position (for example, sudden acceleration or deceleration), and errors in determining the magnetic variation for a specific location. A local magnetic variation can be estimated from available computer models, such as Geomagix

(*http://www.interpex.com/magfield.htm*) or GeoMag (*http://www.resurgentsoftware.com/GeoMag.html*), if the latitude and longitude of the site(s) are known. The magnetic variation also can be determined in the field using techniques described in the WinRiver User's Guide (Teledyne RD Instruments, 2003). When using an external boat-velocity reference (such as GPS), compass errors will affect both measured water velocity and discharge. StreamPro ADCPs do not have an internal compass; therefore, they are not affected by compass errors, but they cannot be integrated with a GPS. Analytical assessment of the compass errors shows that the effect of these errors on velocity and discharge is directly proportional to the speed of the boat. Therefore, maintaining a boat speed that is slow, steady, and practical for the site conditions is imperative to accurately measuring water velocity and discharge when using an external boat-velocity reference.

The accuracy of internal compasses in commercially available ADCPs is typically about +/–1 to 2 degrees. Fluxgate compasses can be unusable when deployed with mounts or boats constructed of ferrous metals or substantial electrical fields. Use of external heading references can improve the accuracy of the heading measurement and eliminate problems associated with ferrous metals and electrical fields. Traditionally, an external heading reference was a gyroscope; however, improvements in GPS technology have made GPS-based heading measurements a cost-effective and accurate solution.

Instrument Quality Assurance

Although ADCPs have no moving parts and typically require no calibration, the instruments and associated software and firmware are complex. Quality-assurance procedures will help identify potential instrument problems. The procedures discussed do not check all components of the ADCP but do identify common problems.

Software and Firmware Procedures

Upgrades to both software and firmware associated with ADCPs are common. Many of these upgrades result in minor improvements to the software or firmware and do not substantially affect the quality of discharge measurements made with the instrument. Nevertheless, some software and firmware changes can be major and can appreciably affect discharge-measurement results. ADCP users, therefore, must ensure that the most recent Office of Surface Water-approved software and firmware will be used for data collection and processing. Firmware and software revisions should be tested before being used for routine data collection. Testing of software and firmware often requires data collection in a variety of conditions with a variety of ancillary equipment. This can be difficult and time consuming and often requires coordination between select groups of users. In addition to information available from instrument manufacturers, the USGS provides information regarding software and firmware in technical memorandums, a mailing list, and the OSW hydroacoustics Web page at *http://hydroacoustics.usgs.gov/.*

Before an ADCP is taken to the field, the most recent OSW-approved software should be installed on the primary and any backup field computers. A copy of the software also should be kept on a storage media separate from the field computers, such as a Universal Serial Bus (USB) memory stick, compact disk-read only memory (CD-ROM), or memory card, in the event of damage or loss of the primary field computer.

Instrument Tests

Each ADCP used should be tested (1) when the ADCP is first acquired, (2) after factory repair and prior to any data collection, (3) after firmware or hardware upgrades and prior to any data collection, and (4) at some periodic interval (for example, annually). The purpose of an instrument test is to verify that the ADCP is working properly for making accurate discharge measurements. Various methods for testing ADCP accuracy include tow-tank tests, flume tests, and comparison of ADCP discharge measurements with discharge measurements from some other source, such as conventional current meters. Each of these methods has limitations as discussed by Oberg (2002).

Beam-Alignment Test

A common source of instrument bias is for the beams to be misaligned. The user can evaluate the potential bias caused by beam misalignment by a simple field test for instruments that have an internal compass. (This test could be applied to a StreamPro ADCP, but extra care would be required to eliminate rotation of the instrument during the test.) The beam-alignment test compares the straight-line distance (commonly called the distance made good) measured by bottom tracking to that measured by GPS. Detailed procedures for the beam-alignment test are provided in Appendix D. Bottom tracking is known to have a small bias caused by terrain effects, but this bias typically is less than 0.2 percent. The USGS-recommended criterion for the Rio Grande ADCP beam alignment to be acceptable is for the ratio of bottom-track distance made good to the GPS distance made good to be between 0.995 and 1.003. For other ADCPs, sufficient data have not been collected to validate this criterion; however, the criterion is assumed to be applicable for other ADCPs. If the instrument does not meet the beam-alignment criterion, the ADCP can be returned to the manufacturer for a custom transformation matrix to be determined and loaded into the instrument.

Periodic Instrument Check

Periodic instrument checks help ensure consistency among instruments and discharge-measurement techniques. The instrument check may be made at a site where the ADCP-measured discharge can be compared with a known discharge derived from some other source, such as the rating discharge from a site with a stable stage-discharge rating or a concurrent measurement made using an independent technique. If the ADCP is equipped with more than one water- or bottom-tracking mode, it is desirable, though not required, to periodically conduct instrument checks by using the different modes. Periodic instrument checks should be performed at different sites, so that a range of hydrologic conditions are reflected in the tests and so that any inherent biases associated with a particular site are minimized. The discharge obtained from the ADCP should be within 5 percent of the known discharge, but a consistent bias in the annual records should be investigated. If the comparison reference is a stable stage-discharge rating and the ADCP measurement departs from the discharge rating by more than 5 percent, it is possible that a rating may have shifted. Another measurement with a second ADCP or conventional discharge measurement should be made to check the validity of the rating before drawing definitive conclusions regarding the ADCP instrument test.

Ancillary Equipment

Although the ADCP and computer are the primary equipment, the ancillary equipment discussed in this section will help achieve an accurate measurement in a variety of conditions. Not all of the equipment discussed is necessary for every measurement; depending on the site conditions encountered, the appropriate equipment should be available.

GPS Requirements and Specifications

Using a GPS to measure the boat velocity is the preferred method of data collection when moving-bed conditions are present. (See Appendix B for a detailed discussion of methods for collecting data in moving-bed conditions.) GPS provides two options for determining boat-velocity differentiated position using the GGA National Marine Electronics Association (NMEA)-0183 sentence or the velocity reported in the VTG NMEA-0183 sentence (National Marine Electronics Association, 2002). In addition to positions, the GGA sentence provides time, elevation, and information about the satellite constellation used to reach the position solutions. The GPS receiver should be configured to output both GGA and VTG sentences.

The boat velocity is determined from the positions in the GGA sentence by dividing the distance between successive positions by the time elapsed between these position solutions (differentiated position). Use of differentiated position requires accurate position solutions and, thus, differential correction. Differential correction compensates for satellite and receiver clock drift, ephemeris inaccuracies, and tropospheric and ionospheric errors associated with the coded signal being broadcast by the GPS satellites and receivers. The two common methods of differentially correcting a GPS signal are (1) real-time kinematic (RTK) systems, which require a user-operated base station and separate rover receiver, both of which can receive dual frequency code-phase and carrier-phase satellite signals, and (2) code-phase differential corrections. RTK systems typically cost tens of thousands of dollars and deliver accuracies in the centimeter range. These systems are used most frequently where satellite-based code-phase corrections are not available or where high-accuracy positions are required. Code-phase differential corrections can be obtained from user-operated base stations but are more commonly obtained from differential correction services. Two free sources of differential correction services are provided by the U.S. Government. The first source is the Wide Area Augmentation System (WAAS), which has been developed for the Federal Aviation Administration (FAA) to provide precision guidance to aircraft at airports and airstrips. The WAAS uses a system of satellites and ground stations that provide GPS signal corrections (Federal Aviation Administration, 2006). The second source is U.S. Coast Guard radio beacons, which are part of a large network that provides differential correction to coastal areas, navigable rivers, and, more recently, inland agricultural areas. Some commercial satellite differential service providers offer differential corrections with various levels of accuracy for a fee. These corrections typically are broadcast using a communications satellite. The accuracy of code-phase differential corrections varies according to the correction source used and the characteristics of the GPS receiver. Many commercial receivers claim submeter accuracy using WAAS as the differential correction source. These receivers are the most common type of receiver used for ADCP data collection and range in cost from about $800 to $3,000.

The velocity reported in the VTG sentence typically is based on measured Doppler shifts in the satellite signals. Use of the Doppler shifts to determine velocity does not require and is unaffected by differential corrections. This velocity measurement can be robust because it is resistant to some of the errors that are problematic for position determination, such as multipath errors and ionospheric and atmospheric distortions. Some receivers, particularly low-cost GPS receivers, may apply filters to smooth out the velocity or display a zero velocity when the velocity drops below a specified threshold. These types of filters and thresholds are unacceptable for using the GPS receiver with an ADCP.

Experience using GPS with ADCPs has shown that the two most common problems are filters in the receiver and multipath errors caused by site conditions. The GPS receiver should allow all filters to be turned off. Multipath errors are caused by the satellite signal reflecting off of bridges, trees, and buildings before arriving at the antenna. Multipath errors affect only the position solutions; they do not affect Doppler-based GPS velocity data. Some receivers contain special antennas or software to reduce multipath errors. If multipath errors are a problem during measurements, the use of VTG often is the best solution. (For additional details about the use of GPS with ADCPs, see Appendix B.)

Echo Sounder

Streams with high sediment concentrations of fine material and sand being transported on or near the streambed also may cause inaccuracies in ADCP water-depth measurements and, therefore, may cause an inaccurate discharge measurement. In such conditions, using a lower frequency echo sounder (approximately 200 kHz) to measure the water depth may be necessary. The echo sounder must support the NMEA 0183 depth below transducer (DBT) data string to be compatible with ADCP data-collection software. If a depth sounder is used, the echo sounder needs to be properly calibrated as part of the premeasurement field procedures. For proper calibration techniques, the user is referred to the bar check procedures in the U.S. Army Corps of Engineers Engineering Design Manual on hydrographic surveying (U.S. Army Corps of Engineers, 2002). The bed-load transport rate and sediment concentration that make the use of a depth sounder necessary have not been quantified, and user judgment is required.

Instrument Deployments and Mounts

Every measurement site has unique features that may determine the best type of ADCP deployment platform. Site features may include hydraulic characteristics, such as water velocity, and access considerations, such as the presence of boat ramps, bridges, or cableways. Three common types of ADCP deployment platforms are manned boats, tethered boats, and remote-controlled boats.

Manned Boats

An ADCP can be mounted on either side, off the bow, or in a well through the hull of a manned boat. Advantages and disadvantages for mounting locations on manned boats are listed in table 4. The ADCP should not be mounted close to any object containing ferrous metal or sources of strong electromagnetic fields, such as generators, batteries, and boat

Table 4. Advantages and disadvantages of acoustic Doppler current profiler (ADCP) mounting locations on manned boats (adapted from Oberg and others, 2005).

Mounting location	Advantages	Disadvantages
Side of boat	Easy to deploy Mounts are easy to construct and are adaptable to a variety of boats ADCP draft measurement can be easily obtained	Moderate chance of directional bias in measured discharges with some boats and flows Possibly closer to ferrous metal (engines) or other sources of electromagnetic fields (EMF) Moderate-low risk of damage to ADCP from debris or obstructions in the water Susceptible to roll-induced bias in ADCP depths
Bow of boat	Minimizes chance of directional bias in measured discharges Mounts are relatively easy to construct Usually far from ferrous metal or electromagnetic fields	Increased risk of damage to ADCP from debris or obstructions in the water More difficult to measure ADCP depth Susceptible to pitch-induced bias in ADCP depths, particularly at high speeds or during rough conditions (waves)
Well in center of boat	Protected from debris and obstructions Accurate depth measurements possible Least susceptible to pitch-and-roll-induced bias in ADCP depths	Often requires special modifications to boat

engines, to minimize ADCP compass errors. A good rule of thumb is that an ADCP should not be mounted any closer to a steel object than the largest dimension of that object. This is a general rule, however, and large variations in the magnetic fields are generated by different metals. Even stainless steel varies appreciably in the amount of ferrous material contained in the steel.

ADCP mounts for manned boats should

- allow the ADCP transducers to be positioned free and clear of the boat hull and mount,

- hold the ADCP in a fixed, vertical position so that the transducers are submerged at all times while minimizing air entrainment under the transducers,

- allow the user to adjust the ADCP depth easily,

- be rigid enough to withstand the force of water caused by the combined water and boat velocity,

- be constructed of non-ferrous materials,

- be adjustable for boat pitch-and-roll, and

- be equipped with a safety cable to hold the ADCP in the event of a mount failure.

Photographs of a variety of ADCP mounts are available in USGS Open-File Report 01-01 (Simpson, 2002, p. 58–69) or at the USGS hydroacoustics Web page (*http://hydroacoustics. usgs.gov/*).

Tethered Boats

A tethered boat can be defined as a small boat (usually less than 6.5 ft long) attached to a rope, or tether, that can be deployed from a bridge, a fixed cableway, or a temporary bank-operated cableway. The tethered boat should be equipped with an ADCP mount that meets all of the specifications outlined in the previous section on manned boats. The tethered boat also should contain a waterproof enclosure capable of housing a power supply and wireless radio modem for data telemetry. A second wireless radio modem attached to the field computer enables communication between the ADCP and field computer without requiring a direct cable connection. The radio modems should reliably communicate with the ADCP using the ADCP data-acquisition software; have a rugged, waterproof housing; operate on a 12-volt direct current (DC) power supply; and have sufficient data-communication capability to maximize ADCP data throughput. Baud rates should be equal to or greater than 9,600 for the RiverSurveyor ADCP and 38,400 for the Rio Grande ADCP (Rehmel and others, 2002). The StreamPro ADCP operates using Bluetooth with a fixed baud rate of 115,200. Rehmel and others (2002) describe the development of a prototype tethered platform, a project to refine the platform into a commercially available product, and tethered-platform measurement procedures. The StreamPro ADCP comes standard with a tethered boat designed specifically for the StreamPro. Tethered boats for other ADCPs, including boats designed to deploy the StreamPro in velocities greater than 5 ft/s, can be purchased commercially from the ADCP manufacturers or third-party vendors.

Tethered ADCP boats have become a common deployment method (fig. 4). Certain considerations need to be made when making tethered ADCP boat measurements. Tethered boats are used in a variety of settings, but primarily they are used from the downstream side of bridges for convenience.

When the water velocity is slow (usually less than 0.5 ft/s), controlling the tethered boat may become difficult. This lack of control may be exacerbated by wind, which may push the boat in an undesired direction. Boat handling can be improved by attaching a sea anchor to the back side of the boat to increase the effect of the current and its pull on the tether. Make sure that this anchor is far enough behind the boat so as to not disturb the flow and potentially bias the velocity measurements.

When the water velocity is fast (usually greater than 5 ft/s) or when the boat is deployed from a high bridge, it is not uncommon for a tethered boat to be pitched upward at the bow. This increased pitch is caused by increased vertical tension on the tether in faster flows, hull dynamics, and an

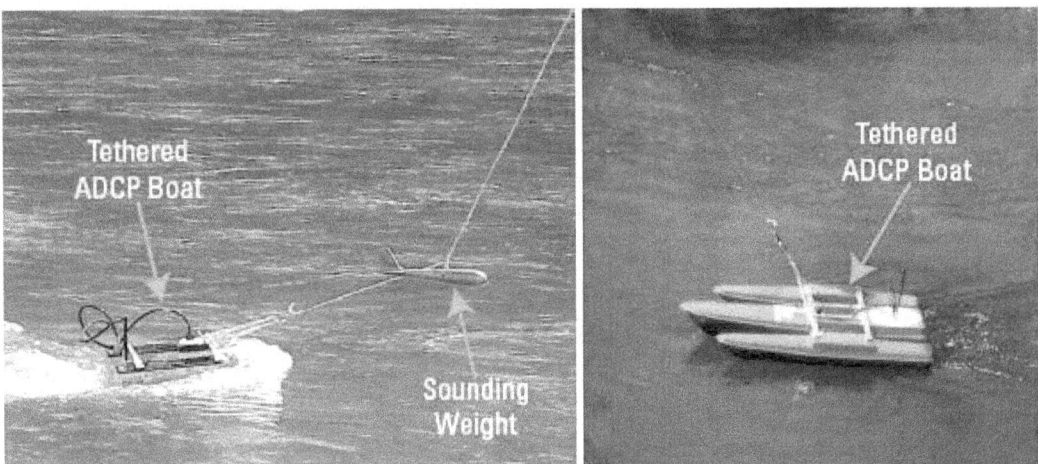

Figure 4.   Examples of tethered acoustic Doppler current profiler (ADCP) boats used for making discharge measurements (left photograph by Jeff Woodward, Environment Canada (used with permission); right photograph by Geoffrey D. Cartano, U.S. Geological Survey).

Bridge piers can cause excessive turbulence during high streamflow, especially if debris accumulations are present on the piers and the piers are skewed to the flow. The effect of bridge-pier-induced turbulence may be reduced by lengthening the tether to increase the distance between the bridge and the tethered boat. Attention should be paid to the cross section to ensure that no large eddies that could cause flow to be nonhomogeneous. Possible alternatives to measuring off the downstream side of bridges include using bank-operated cableways or having personnel on each bank hold a rope attached to the platform to pull the platform back and forth across the river. Bank-operated cableways may be as simple as a temporary "rope and pulley" apparatus (fig. 5) or may involve the use of a small temporary cableway with a motorized drive for towing the tethered boat back and forth across the stream (fig. 6). In 2004, remote-controlled rovers were developed for cableways. These rovers can be carried from one streamflow-gaging station to another and, once mounted on the cableway, can be used to winch up the tethered boat and drive the boat back and forth at a user-controlled speed.

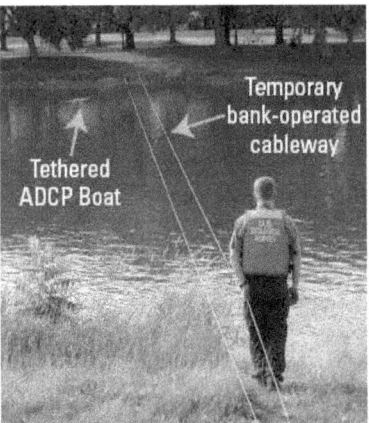

Figure 5.   Temporary bank-operated cableway for making acoustic Doppler current profiler (ADCP) measurements with a tethered ADCP boat (photograph by Brian L. Loving, U.S. Geological Survey).

incorrect setting of the angle for the bail for those boats equipped with a rigid bail. The bail connects the tether to the boat and can be either a rigid design or a flexible rope bail. Large pitch angles may introduce some bias in depth measurements and should be minimized as much as practical. Adding a sounding weight on the tether near the location where the tether is tied to the boat (fig. 4) will help decrease the pitch angle. In addition, increasing the length of the tether helps reduce the pitch angle.

The tether line should be visible from the water surface to minimize the risk of collision with river traffic. Orange plastic flags tied along the tether will enhance its visibility. The operator should also be capable of releasing the tether quickly in case the boat becomes entangled in debris or collides with river traffic. Do not wind the tether around the hand to hold the boat because this is a safety hazard. Standard safety practices, site-specific traffic safety plans, and the local highway traffic regulations should be followed.

It is possible to lose control of a tethered boat because of a system-component failure. For example, a boat tether or tether attachment point could break. ADCP operators using tethered-boat deployments should have redundant attachment points for the tether on the boat and have a contingency plan for retrieving the boat in the event of a failure that causes a loss of boat control. An example of a contingency plan would be to carry a small manned boat that could be quickly and safely launched to retrieve the tethered boat (Oberg and others, 2005).

Remote-Controlled Boats

Unmanned, remote-controlled boats allow the deployment of ADCPs where deployment with a manned boat or tethered boat may not be feasible or ideal. Similar to (but smaller than) manned boats, a remote-controlled boat has self-contained motors and a remote-controlled system for maneuvering the boat across the river. Unlike the tethered boat, the remote-controlled boat has no rope (tether) restraints. Although remote-controlled boats have an increased risk of equipment loss because of potential loss of boat control, they provide the ability to launch a boat without a boat ramp and to collect data away from bridge effects (for example, upstream from a bridge)

Figure 6. Motorized cableway rover for deploying tethered acoustic Doppler current profilers (photograph courtesy of Water Survey of Canada).

or at sites where no bridge or cableway is present. Currently (2008) remote-controlled boats are commercially available, but extensive field experience with these boats has not been obtained (fig. 7).

ADCP mount for a remote-controlled boat should meet all mount specifications previously listed for manned boats. The remote-controlled boat also should contain a waterproof enclosure capable of housing a power supply, a radio modem, and the control radio. Radio modems are used for data telemetry between the remote-controlled boat and field computer; the radio modems should have the capabilities previously described for tethered-boat deployments.

The same operational guidelines regarding speed and maneuvering for manned boats also apply to remote-controlled boats. Proper control of a remote-controlled boat requires practice. The operator should be familiar with remote-controlled boat operation prior to using this deployment

A. SeaRobotics, Inc.

B. OceanScience Group

Figure 7. Examples of commercially available remote-controlled boats (photographs courtesy of SeaRobotics, Inc., and OceanScience Group).

technique in high flows. Regular maintenance of the boat and control radios is critical to ensure reliable operation.

For remote-controlled boats, it is possible to lose control of the boat because of a system component failure. ADCP operators using remote-controlled boat deployments should have a contingency plan for retrieving the boat in the event of a failure that causes a loss of boat control. An example of a contingency plan would be to carry a small manned boat that could be quickly and safely launched to retrieve the remote-controlled boat (Oberg and others, 2005).

Other Equipment

A serial port on a laptop computer is required for communication with the ADCP, and a second serial port is required if a GPS is used. New laptop computers typically do not contain a serial port. Use of USB or Personal Computer Memory Card International Association (PCMCIA) serial ports is often required. USB serial ports are virtual serial ports, and some brands to do not work well with ADCPs and(or) GPS. StreamPro ADCPs communicate with portable digital assistants (PDAs) and laptop computers through a wireless Bluetooth connection. An external Bluetooth radio is required for computers that do not have built-in Bluetooth. Prior to going to the field, all ports should be checked for compatibility with the instruments to be used. A list of some USB, PCMCIA, and Bluetooth serial ports that have been shown to work well with

ADCPs can be found at the USGS hydroacoustics Web site (*http://hydroacoustics.usgs.gov/*).

In addition to the ADCP and computer, the following additional equipment is necessary to achieve a high-quality discharge measurement (table 5).

- A *toolkit* (fig. 8) should be assembled for the ADCP with tools, multimeter, and any spare parts that may be difficult to obtain in the field (such as fuses, o-rings, and special wrenches). The toolkit should always be kept with the ADCP.

- An adequate supply of the *OSW-approved ADCP discharge-measurement field forms* should be taken to the field. The USGS discharge-measurement form (9-275-I) is available from the USGS hydroacoustics Web site (*http://hydroacoustics.usgs.gov/*) and is shown in Appendix E, figure E-2.

- *Computer data-storage media* (such as a flash-memory card, USB memory stick, or CD-ROM) should be available with sufficient storage space for making temporary backup copies of all field data files.

- A *thermometer* is needed to check the accuracy of the water-temperature measurement of the ADCP because an incorrect temperature will bias the velocity and discharge measurements.

Table 5. List of ancillary equipment to be used with acoustic Doppler current profilers when making streamflow measurements.

[USB, Universal Serial Bus; PCMCIA, Personal Computer Memory Card International Association; ADCP, acoustic Doppler current profiler; DGPS, differentially corrected global positioning system]

Equipment	Function	Optional or required
USB or PCMCIA serial port(s)	Computer connection to ADCP and(or) DGPS	Optional[a]
External Bluetooth radio	Computer connection to StreamPro	Optional[b]
Toolkit	Field troubleshooting and repairs	Required
ADCP field notes	Note keeping	Required
Computer data-storage media	Field backups of data	Required
Thermometer	Measure water and air temperatures	Required
Measuring tape, or mount with graduated markings, or folding rule	Measure ADCP depth	Required
Laser rangefinder or other distance-measurement tool	Measure shore distances	Required
Salinity/conductivity meter	Measure salinity	Optional[c]
Hand-held anemometer	Obtain estimate of wind speed	Optional
ADCP cable	Direct communication	Required
Trolling motor or tagline	Slow boat speed	Optional
Hand-held radios	Communication during tethered or remote-controlled boat measurements	Optional
Shade/rain cover for computer	Improves view of computer screen/protects screen	Optional

[a] Required if computer does not support sufficient internal serial ports.

[b] Required if computer does not support internal Bluetooth communication.

[c] Required for ADCP measurements in estuaries and coastal streams.

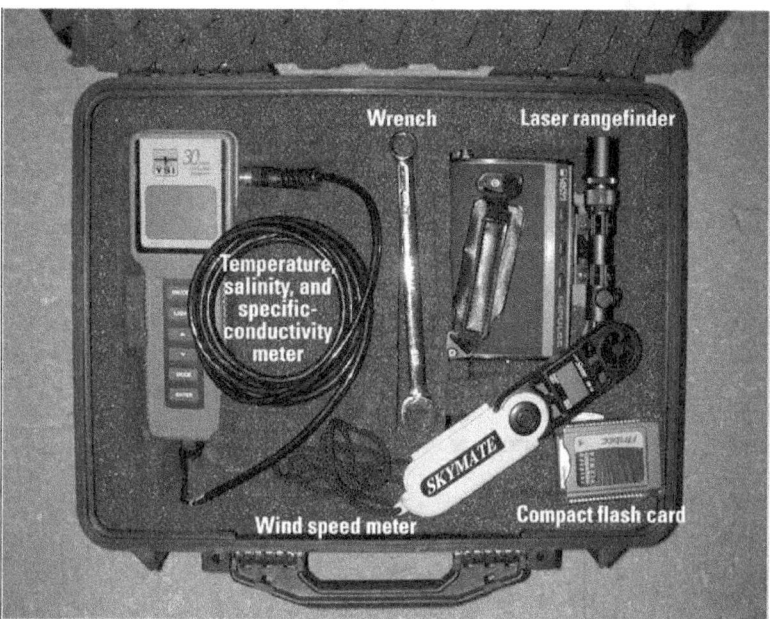

Figure 8. Example toolkit of ancillary equipment for use with acoustic Doppler current profilers (ADCPs) when making streamflow measurements (photograph by John M. Shelton, U.S. Geological Survey).

- If the ADCP mount does not have *graduated markings*, a *measuring tape* or *folding rule* is needed to measure the depth of the ADCP.

- The distances from the beginning and end of a transect to the nearest edge of water must be measured and input to the software for computation of the discharge in the unmeasured areas. Typically visual estimates underestimate the distance over water; therefore, a *laser rangefinder* or some other means of measuring the distance to shore is required. The calibration of distance-measurement devices should be checked periodically by measuring the distance to targets at a known distance; the results of these calibration tests should be recorded in an office log. Various types of laser and optical rangefinders, accuracy and limitations, and test results can be found on the USGS hydroacoustics Web site (*http://hydroacoustics.usgs.gov/*).

- A *conductivity* or *salinity meter* is required to determine the salinity at the transducer face for measurement in saline environments.

- If the surface velocities are affected by wind, a *hand-held anemometer* will allow accurate characterization of the wind speed and direction.

- If radio modems are used for ADCP communications, the *cable* for connecting directly to the ADCP should be taken to the field. An ADCP connected through radio modems occasionally will not communicate with the field computer. The problem is often resolved by using a direct cable to establish communications and "reset" the ADCP for modem use. If a second pair of radio modems is available, they should be taken to the field as a backup.

- If low-flow conditions are expected (generally velocities less than 0.5 ft/s), a *trolling motor* or *tagline* may be necessary to keep boat speed slow and consistent (Oberg and others, 2005). If it is not possible to maintain a slow boat speed, maintain the slowest speed that allows smooth boat operation. (Additional transects may be necessary to average turbulence and instrument noise.)

- If a remote or tethered boat deployment is used, *hand-held radios* are helpful for communications between the boat operator and the computer operator.

- If the data-collection computer will be used in bright sunlight, a *shade for the computer screen* may be necessary to improve the readability of the screen. In potentially rainy conditions, a *rain cover* is required to keep non-ruggedized computers dry.

Final Equipment Preparation and Inspection

A pre-field inspection checklist is recommended to ensure that all procedures are followed and that all necessary equipment is available and functioning for the field trip. An example of a pre-field inspection checklist is shown in Appendix E, figure E-1; however, the checklist should only be used as a guideline for field preparation. Other equipment may be necessary for the sites and conditions that may be encountered in the field. The ADCP, cables, connectors, batteries, mounts, and GPS or echo sounders that will be integrated with the ADCP in the field should be inspected for any irregularities. The ADCP should be connected to the field computer, and communications with the ADCP should be established using the ADCP data-collection software and computer to be used in the field. The ADCP clock should then be set to the appropriate reference time (usually local or Greenwich Mean Time). If radio modems are to be used for communications with a tethered or remote-controlled boat, the communications should be established using the radio modems. If a GPS or echo sounder will be connected to the ADCP in the field, then the GPS or echo sounder should be connected with the ADCP to the computer to ensure that they properly function with the ADCP and the ADCP data-collection software. If problems are encountered during any system check, the problems should be resolved by (1) consulting the necessary technical documentation, (2) calling a qualified agency staff member familiar with ADCPs, (3) calling the vendor technical support unit, or some combination of these three options.

Tethered and remote-controlled boat hulls, fins, structural members and compartments/hatches should be inspected. The tether line(s) and connectors for a tethered boat should be inspected for wear and to ensure that they are suitable to withstand expected field conditions. When deployed in streams and rivers with high velocities or turbulence, redundant attachment points for the tether on the boat should be used to allow the tethered boat to be recovered should the primary attachment point fail. Remote-controlled boat deployments, motors, servos, and the radio controller should be inspected and tested before going into the field (Oberg and others, 2005).

Field Procedures

Proper field procedures are critical to obtaining high-quality discharge measurements using ADCPs. Although step-by-step procedures are an important aspect of high-quality data collection, nothing can substitute for field personnel who understand both the instrument and effect of hydraulic and sediment transport conditions. The following sections provide guidance on proper site selection, pre-measurement field procedures, discharge-measurement procedures, and post-measurement field procedures.

Site Selection

One of the most important steps in collecting high-quality streamflow measurements is site selection. Many ADCP measurement problems can be solved by moving to a better measurement site. The guidelines provided in USGS Water-Supply Paper 2175 (Rantz and others, 1982, p. 139) for traditional current-meter measurements also are excellent guidelines when using an ADCP, except for those guidelines that relate to depth and velocity requirements for specific meters. General guidelines for selection of an ADCP measurement site can be categorized by location, shape, flow velocity, and other factors.

1. Location:

   a. The cross section of a stream lies within a straight reach, and streamlines are parallel to each other. Flow is relatively uniform and free of eddies, slack water, and excessive turbulence (Rantz and others, 1982).

   b. The measurement section is relatively close to the gaging station control to avoid the effect of tributary inflow between the measurement section and control and to avoid the effect of storage between the measurement section and control during periods of rapidly changing stage (Rantz and others, 1982).

2. Shape:

   a. Desirable measurement sections are roughly parabolic, trapezoidal, or rectangular. Asymmetric channel geometries (for example, deep on one side and shallow on the other) should be avoided if possible (Simpson, 2002), as should cross sections with abrupt changes in channel-bottom slope.

   b. The streambed cross section should be as uniform as possible and free from debris and vegetation or plant growth.

   c. Depth at the measurement site should allow for the measurement of velocity in two or more depth cells at the start and stop points near the left and right edges of the measurement section.

3. Flow velocity:

   a. Measurement sections with mean velocities less than 0.3 ft/s should be avoided if an alternative measurement location is available (Oberg and others, 2005). Although measurements can be made in low velocities, boat speeds must be kept extremely slow (if possible, less than or equal to the average water velocity) requiring special techniques for boat control (Simpson, 2002). If maintaining a slow boat speed is not possible, maintain the slowest speed that allows smooth boat operation. (Additional transects may be necessary to average turbulence and instrument noise.)

    b. Sites with very turbulent flow, such as standing waves, large eddies, and non-uniform flow lines, should be avoided. This condition is often indicative of non-homogenous flow, which is a condition that violates one of the assumptions required for accurate ADCP velocity and discharge measurements.

  4. Other factors:

    a. Measurement sections having local magnetic fields that are relatively large as compared to the Earth's magnetic field should be avoided. Large steel structures, such as overhead truss bridges, are a common source for these large local magnetic fields and may result in ADCP compass errors.

    b. When using GPS, avoid locations where multipath interference is possible, such as where signals from the satellites bounce off structures and objects such as trees along the bank or nearby bridges or buildings. Also avoid locations where reception of signals from GPS satellites is blocked. It may be possible to make valid measurements in sections that violate one or more of the above guidelines, but whenever possible, locate and use a better measurement section (Oberg and others, 2005).

Sometimes measurements need to be made where conditions do not satisfy the suggested guidelines. In such situations, the quality of the measurement can be greatly diminished; therefore, the field personnel must use their best judgment in selecting a measurement section.

Pre-Measurement Field Procedures

Pre-measurement tests and proper configuration of the ADCP help to ensure a high-quality measurement. The following sections describe the field procedures that should be completed prior to ADCP measurements.

Set Internal Clock

Prior to the start of the discharge measurement, the ADCP's internal clock should be checked, set to the correct time, and noted on the ADCP measurement field form. This should be done prior to any diagnostic tests, calibrations, or

configuration so that time stamps on all data will be consistent. In most cases, the ADCP's clock should be set to agree with the recorder time at the streamflow-gaging station. Checking and setting the correct time is of particular importance when using the discharge measurements to calibrate or check the calibration of fixed acoustic current meters installed at streamflow-gaging stations, or when measuring at sites where the flow is unsteady (Oberg and others, 2005).

Instrument Diagnostic Checks

After the ADCP is mounted and communication between the ADCP and field computer is established, the ADCP must be checked to ensure all components are operating properly. Diagnostic tests should be performed and the results electronically stored on the field computer. Diagnostic tests should include system serial number, firmware and hardware configuration versions, the beam transformation matrix, electronics diagnostic tests, internal system tests, and sensor verification tests. Software for executing diagnostic tests for some ADCPs is available on the USGS hydroacoustics Web site or built into the data-collection software available from the manufacturer. If software or diagnostic-test information is not available for a specific instrument, the user should contact the manufacturer for guidance. The results of diagnostic tests should be backed up in the field and archived in the office with the associated discharge-measurement files. Diagnostics tests should be documented on the ADCP discharge-measurement field form (Appendix E, figure E-2). Complete diagnostic tests should be made prior to every discharge measurement. If possible, conduct the test from a stationary boat in relatively still water, for example, near the shore. Some of the tests require little or no water motion relative to the ADCP.

Speed of Sound

Variation in the speed of sound with depth does not affect the measurement of horizontal currents (Teledyne RD Instruments, 1996); however, it does affect the measurement of vertical currents and depth (range from the transducer). Regarding horizontal velocity measurements, Snell's law states that the horizontal wave number is conserved when sound passes through horizontal interfaces. Because the frequency of the sound wave remains constant, the change in the speed of sound with depth does not affect the horizontal component of the sound velocity and hence the horizontal velocity measurement. The measurement of the vertical velocity component and the depth is proportional to the change in speed of sound. Currently (2008), commercially available data-collection and processing software do not have the capability for correcting the vertical velocity or the depth for changes in the speed of sound. Temperature and salinity are the two most important variables for determining the speed of sound for boat-mounted ADCPs.

Water Temperature

ADCPs have built-in temperature sensors to measure water temperature at the transducer face. Temperature is the most important variable in the equation used to compute the speed of sound (Urick, 1983, p. 113). The ADCP must compute the speed of sound correctly to accurately measure velocities, depths, and compute discharge. An error of 5 degrees Celsius (°C) in the temperature measurement will cause a 2-percent bias error in the measured discharge (Oberg and others, 2005). Thus, the temperature measured by the ADCP should be compared with an independent temperature measurement made adjacent to the ADCP. This check should be performed prior to every discharge measurement and the results recorded on the measurement field form. If the temperature measured by the ADCP temperature sensor differs consistently from the independent temperature measurement by 2 °C or more, or if the ADCP temperature sensor has failed, the ADCP should not be used to make discharge measurements until the temperature sensor is repaired and checked. In the event that a discharge measurement is necessary and another ADCP is not readily available, it may be possible to enter a temperature manually for use in the speed-of-sound calculations. This action is not recommended as standard practice, however, and it may decrease the accuracy of the discharge measurement.

Salinity

Salinity is another important variable in the speed of sound equation. Salinity values generally range from zero parts per thousand (ppt) for freshwater to 35 ppt for water from the open ocean. When measuring in waters where the salinity is greater than zero, the salinity should be measured near the transducer face and recorded on the field form. The salinity value may then be entered into the ADCP data-collection software prior to data collection and adjusted as necessary during measurement playback and processing. Salinity should be measured for every transect in locations where salinity varies over time. Salinity may also vary from bank to bank. It should be noted that the salinity value used for a transect should reflect an average salinity for the section to be measured at the approximate depth of the ADCP transducers.

Compass Calibration

Calibrating the internal magnetic compass of instruments with an internal compass is encouraged prior to all ADCP measurements, but is mandatory when using GPS as the navigation reference, using the loop method for correcting discharge for bias caused by a moving bed (Mueller and Wagner, 2006), or when the velocity direction is important. The instrument-specific procedures available from the manufacturer for calibrating the compass of the ADCP being used should be followed. An automated program from some

ADCPs is available on the USGS hydroacoustics Web site (*http://hydroacoustics.usgs.gov/*). The following guidelines should be followed for successful compass calibrations:

1. Minimize ferrous material and electromagnetic field (EMF) interference located in the vicinity of the ADCP (on the boat and at the measurement site). EMF and ferrous material may adversely affect the performance of the internal magnetic compass.

2. If the instrument calibration or evaluation process reports a total compass error, this error should be less than 1 degree when evaluated after calibration. If the error reported by the compass evaluation exceeds 1 degree, the calibration procedure should be repeated. If after several attempts, the total compass error cannot be reduced to less than 1 degree, the compass error should be noted on the field sheet. The discharge measurement then can be made, but special attention should be paid to potential compass errors, such as directional bias and irregular ship track. If the instrument does not report a numerical error, the instrument should report an "excellent" rating for the horizontal calibration parameter.

3. Pitch-and-roll changes must be minimized with Rio Grande ADCPs because they are equipped only for a single-tilt calibration process. In such situations, the standard deviation of the pitch-and-roll should be less than 1 degree and, ideally, 0.5 degree or less during the calibration and evaluation process. For RiverSurveyor ADCPs, pitching and rolling the instrument is encouraged during the calibration, but only the rating of the horizontal calibration is of major concern.

4. Compass calibration should be done as close to the measurement site as possible with the ADCP mounted in the same manner as it will be deployed for the measurement.

5. For best results, the maximum rotation velocity should be 5 degrees per second or less.

Instrument Configuration

The ADCP should be configured by a trained user to reflect the hydrologic conditions at the site and to optimize the data quality (Lipscomb, 1995). ADCP configuration parameters that must be set include the blanking distance, water mode (if applicable), depth-cell size, and profiling range. Other parameters that should be set prior to data collection but can be modified during postprocessing include the instrument draft, edge shape, top and bottom extrapolation method, and magnetic variation. Configuration parameters are specific to the type (narrowband or broadband), the manufacturer, and the model of the ADCP being used. For a detailed description of all configuration parameters, refer to the technical documentation for the specific ADCP. General recommendations for configuration parameters are given below.

1. File names for the data files collected (also called deployment names) should follow a uniform, documented convention developed by each office involved in ADCP operation (U.S. Geological Survey, 2005a).

2. The depth of the ADCP (vertical distance from the water surface to the center of the transducer face) must be measured accurately, recorded in the ADCP discharge-measurement notes, and entered into the configuration file. The pitch-and-roll of the boat when the depth is measured should be similar to the pitch-and-roll during the discharge measurement. If the depth of the ADCP changes during the measurement, the depth must be measured again, noted, and the configuration file modified with the new depth.

3. Most ADCP data-collection software contains an automated method to configure the ADCP. The automated methods are dependent upon user-supplied information about site characteristics, such as maximum water depth, bed-material characteristics, and expected maximum water and boat speeds. Where these methods are available in ADCP data-collection software, the software should be used to configure the ADCP for discharge measurements (U.S. Geological Survey, 2003). The commands generated by this software utility, however, should be checked prior to the start of the measurement.

4. The configuration parameters and the site conditions entered into an automated configuration program should be documented in the field notes. Changes made to the ADCP configuration during a measurement should be documented on the measurement field note forms, so it is clear that changes were made and to which transects these changes apply.

5. Manual configuration of an ADCP should only be used in rare cases where the automated procedures are not applicable. The most up-to-date agency-specific guidelines for the instrument should be understood before attempting a manual configuration. If guidelines are not available, the user should use manufacturer recommendations for the unit. Also, seek advice from USGS ADCP experts.

6. Configuration of the ADCP to collect single-ping water data is preferable, if random noise levels do not prohibit this configuration. Collection of single-ping data allows possible data-quality problems to be more easily identified than problems with multi-ping averaged data. When collecting multi-ping averaged data, the user should be aware of how often the

heading, pitch, and roll sensors are recorded and how often water depth and boat velocity are measured. Typically, this is done automatically in most narrow-band ADCPs, but the flexibility provided by water mode 12 in broadband ADCPs allows the user to set a configuration that is not optimal for moving-boat deployments. If the averaging interval is too long for the boat stability and water turbulence, errors can be introduced into the measurement.

7. The extrapolation method for the top and bottom unmeasured zones must be specified unless data are collected with a StreamPro ADCP on a PDA, in which case, the extrapolation methods default to the one-sixth (0.1667 power coefficient) power law on the top and bottom for data collection. Often, the appropriate extrapolation method cannot be determined until after the measurement during post-processing. Previous data collected at a site may be used to guide the selection of the extrapolation method. In the absence of any other information, the one-sixth power-law extrapolation method is a good technique for most open-water discharge measurements made during steady-flow conditions. The extrapolation methods should be evaluated and, if necessary, changed during post-processing.

8. Wind speed can be important, especially for sites with low velocities where wind can greatly affect the surface velocities and influence the top extrapolation method to be applied. Overall wind speed and direction, as well as changes between transects, should be noted on all measurement field note forms to assist with accurate processing and reviewing of measurements.

9. If the user is unfamiliar with the measurement section, a trial transect, which may or may not be recorded, should be made across the river. A trial transect is useful for determining the following characteristics of the proposed measurement:

    a. maximum water depth;

    b. overall cross-section shape;

    c. maximum water velocity and its location in the cross section;

    d. flow uniformity;

    e. effects of hydraulic structures, such as bridges, piers, and islands, on the flow;

    f. unusual flow conditions, such as reverse or bi-directional flow;

    g. bank shapes; and

h. approximate start-and-stop locations on the left and right banks, where a minimum of two depth cells with valid velocity measurements can be measured. (To obtain consistent edge estimates, buoys can be used to mark the start-and-stop locations.)

The information gleaned from the trial transect should be recorded on the discharge-measurement notes.

Moving-Bed Tests

ADCPs can measure boat velocity using a technique called bottom tracking, which computes the Doppler shift of acoustic pulses reflected from the streambed. This technique assumes that the streambed is stationary; however, sediment transport on or near the streambed can affect the Doppler shift of the bottom-tracking pulses. In such situations, reflections of bottom-tracking pulses from highly concentrated near-bed sediments contaminate the reflections from the bed. These near-bed sediments typically are being transported in the downstream direction. If bottom tracking is affected by sediment transport, the measured boat velocity will be biased in the opposite direction of the sediment movement. A stationary boat in the stream would appear to be moving upstream (fig. 9). This bias in the boat velocity will result in measured velocities and discharge that are less than the true velocities and discharge (negative bias).

USGS policy (U.S. Geological Survey, 2002b; Oberg and others, 2005) requires that a moving-bed test be conducted prior to making a discharge measurement. In USGS training classes and in Oberg and others (2005), an exception to conducting a moving-bed test with every measurement was allowed if sufficient documentation was provided in the field folder to justify not conducting a moving-bed test. Field experience has shown that sediment transport characteristics can vary greatly for the same discharge, depending on the hydrograph shape, source of runoff, and season of the year. Thus, a moving bed may be detected at a location and discharge where one was not previously detected, or a moving bed may no longer exist at a location and discharge where one was previously detected. Moving-bed conditions also have been observed in low-velocity environments (less than 1 ft/s) and are likely caused by organic material transported by the water. Therefore, to ensure the quality of the data collected, every moving-boat measurement made with an ADCP must have a recorded moving-bed test. If a site routinely has a moving bed and GPS is always used with the ADCP, a moving-bed test is still required but need only be 5 minutes in length. This requirement supersedes Oberg and others (2005) and various USGS training materials.

A moving-bed test is only useful if proper techniques are followed in conducting the test and analyzing the results. The three acceptable methods for performing a moving-bed test are (1) stationary test with no GPS, (2) stationary test with GPS, and (3) the loop method. A brief summary of each method is provided here, and detailed descriptions of these methods are provided in Appendix B along with data-collection and processing methods that can be used to obtain an unbiased discharge measurement if a moving bed is present.

The stationary test with no GPS requires that the boat with the ADCP be held in a stationary position while recording ADCP data, using bottom tracking as the boat-velocity reference. If the stationary position is maintained by a tether or anchor so that upstream or downstream movement of the ADCP is not possible, the moving-bed test should be recorded for no less than 5 minutes; however, if the ADCP can move either upstream or downstream, such as when the boat operator is trying to maintain position of the boat, the test should be recorded for no less than 10 minutes. These criteria supersede the guidance on stationary moving-bed tests that have been published previously in U.S. Geological Survey (2002b) and Oberg and others (2005). When a moving-bed condition is present, a stationary boat will appear to have moved upstream (fig. 9). The error caused by the moving bed can be estimated by dividing the distance of the apparent boat motion in the upstream direction by the duration of the test in seconds. This computation will provide an estimate of the moving bed detected by the bottom-tracking technique. This moving-bed velocity can then be divided by the average water velocity and the moving-bed velocity, and multiplied by 100 to yield the percent bias error for a water-velocity measurement at this stream location. If the moving-bed test was completed with a fixed tethered deployment, an anchored

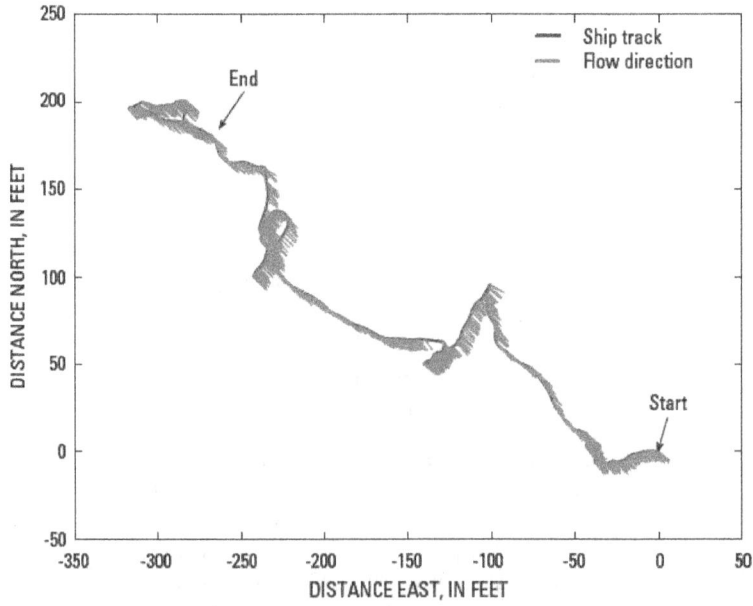

Figure 9. Example of a moving bed measured with a 1,200-kilohertz acoustic Doppler current profiler (ADCP) on the Mississippi River at Chester, Illinois.

manned boat, or a manned boat where little movement of the boat was ensured, a moving bed is determined to be present when the measured moving-bed velocity is greater than 1 percent of the mean water velocity at the test location. If the moving-bed test was conducted using a manned boat that was not anchored and may have moved either upstream or downstream, a criteria of 2 percent instead of 1 percent is used because uncertainty has been introduced into the test by the boat's movement. Discharge-measurement techniques that are not affected by a moving bed, or that correct for the effect of a moving bed, should be used if a moving bed has been detected (Appendix B).

A more accurate method for estimating the errors introduced by a moving bed can be determined if a GPS is available for use and is interfaced with the ADCP and the data-collection software. This second method also requires that the ADCP boat be held in a stationary position and a data file recorded for at least 5 minutes, if quality GPS data are being recorded. The error caused by the moving bed can be computed in the same manner as previously described for the first method, except that the distance in the upstream direction indicated by bottom tracking should be corrected by the distance actually traveled in that direction, as indicated by GPS (Oberg and others, 2005). In the WinRiver software, this distance can be found in the "compass calibration" tabular window and is labeled "BMG-GMG mag," and the direction of the "BMG-GMG dir" should be in the upstream direction. If the measured moving-bed velocity is greater than 1 percent of the mean water velocity at the test location, discharge-measurement techniques that are not affected by a moving bed, or that correct for the effect of a moving bed, should be used (Appendix B).

If the ADCP can be held stationary, stationary moving-bed tests are a good measure of the magnitude of an apparent moving streambed; however, these tests represent moving-bed conditions for only one location in the cross section. An alternative to the stationary moving-bed test is the loop method, which is based on the fact that as an ADCP is moved across the stream, a moving bed will cause the bottom-track-based ship track to be distorted in the upstream direction. Therefore, if an ADCP makes a two-way crossing of a stream (loop) with a moving bed and returns to the exact starting position, the bottom-track-based ship track will show that the ADCP appears to have returned to a position upstream from the original starting position (fig. 10). The mean moving-bed velocity can be computed from the distance the ADCP appeared to have moved upstream from the starting position (loop-closure error) and the time required to complete the loop. If the moving-bed velocity measured by the loop method is greater than 0.04 ft/s and greater than 1 percent of the mean water velocity, a moving bed is present. Discharge-measurement techniques that are not affected by a moving bed, or that correct for the effect of a moving bed, should be used if a moving bed has been detected (Appendix B). The loop method must be applied properly, or it may produce incorrect results. Anyone planning to use the loop method should read

and follow USGS Scientific Investigations Report 2006–5079 (Mueller and Wagner, 2006), which describes the procedures, limitations, and uncertainties associated with the loop method. A detailed description of the loop method also is presented in Appendix B.

Discharge-Measurement Procedures

The procedures to be followed to make quality discharge measurements vary depending on the flow conditions being measured. The procedures for measuring steady-flow conditions are different from the procedures used to measure unsteady-flow conditions. Although the procedures may be different for the various flow conditions, the data-quality indicators for both conditions are consistent. The following sections provide details on the recommended procedures for measuring discharge in steady- and unsteady-flow conditions as well as data-quality problems to monitor in the field when making discharge measurements.

Steady-Flow Conditions

A discharge measurement in steady-flow conditions is obtained from the measurement of a minimum of four transects (two in each direction). The measured discharge is the average of the discharges from the four transects. If the discharge for any of the four transects differs by more than 5 percent from the mean measured discharge and no critical data-quality problem can be identified and documented, a minimum of four additional transects should be obtained, and the mean of all eight transects will be the measured discharge (U.S. Geological Survey, 2002b). If the discharge for one or more transects is not within 5 percent of the mean measured discharge and a critical data-quality problem can be identified

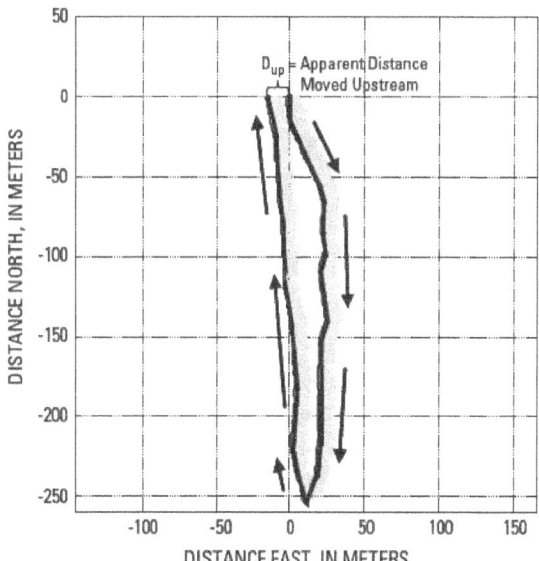

Figure 10. A distorted ship track in a loop caused by a moving bed.

and documented (for example, a tow boat approaching the section, a sudden change in discharge because of a lockage, communication problems between the computer and ADCP, or other factors), the transect deviating from the mean may be replaced with an additional transect collected in the same direction. Reciprocal transects should always be measured to reduce potential directional biases. Directional biases occur when the discharges measured for transects from the left bank to the right bank are consistently either greater than or less than discharges measured for transects made from the right bank to the left bank.

When the mean channel velocity is less than 0.8 ft/s, the TRDI StreamPro ADCP discharge measurements for individual transects have much greater variability than those StreamPro measurements made when the mean channel velocity is greater than 0.8 ft/s. Discharge measurements made when mean velocities were less than 0.8 ft/s had an average coefficient of variation for individual transect discharges of 12 percent, whereas measurements with mean velocities greater than 0.8 ft/s had an average coefficient of variation of 2.5 percent. Despite this larger variation, the measured discharges (the mean discharge for all transects) do not seem to be biased, provided that enough transects (potentially more than eight) are included in the mean discharge. When using a StreamPro ADCP in these slow conditions, a slow, steady boat speed is critical, and water mode 13 (WM13) should be used if the site conditions meet the criteria for maximum water speed (less than 0.8 ft/s) and depth (less than 3.3 ft). Additional details on the StreamPro ADCP testing results can be found in OSW Technical Memorandum 2005.05 (U.S. Geological Survey, 2005b).

Unsteady-Flow Conditions

At times, flow changes rapidly enough that discharge measurements within 5 percent of the mean cannot be collected from four transects. Unsteady flows can be caused by upstream dam or lock regulation, tidal effects, downstream backwater effects, flood waves, or other conditions. It may be necessary to use measurements from individual transects as discrete measurements of discharge if the flow is changing rapidly. If possible, however, pairs of reciprocal transects should be averaged together as one measurement of discharge to reduce the potential of directional bias (U.S. Geological Survey, 2002b). The justification for using a single transect or pairs of transects for discharge measurements should be documented in the field notes and stored with the discharge measurement or applicable station analysis files. Another consideration for unsteady flows, specifically bi-directional flows, is the assignment of a positive or negative sign to the discharge measurement. The ADCP software may or may not assign flow direction correctly, and the positive or negative sign also can change depending on which edge is designated "left" or "right." Thus, the operator should note the direction of flow during measurement for each transect, according to accepted convention for a particular site.

Critical Data-Quality Problems

When making ADCP discharge measurements, the ADCP operator should continuously monitor the data through the ADCP software. If a critical data-quality problem is observed during measurement at a transect, the use of that transect may be terminated. If a transect is not used, the reason should be documented on the ADCP discharge-measurement field note form, and that transect should not be used in the computation of measurement discharge. If the problem was related to undesirable measurement-section characteristics, a new measurement section should be located and noted on the measurement field note form. If the terminated transect was not the first transect in a measurement series, the boat should be returned to the initial starting point to ensure the transects are measured in reciprocal pairs (Oberg and others, 2005). Potential critical data-quality problems can include, but are not limited to the following:

a. inappropriate or improperly configured water or bottom mode;

b. configuration errors, such as an insufficient number of depth cells to profile down to the channel bed;

c. appreciable or consistent portion of the section with invalid or missing data (invalid data failed to meet internal and user-specified data-quality criteria, and missing data are a result of communication problems between the computer and the ADCP);

d. appreciable invalid bottom tracking;

e. erroneous boat or water velocities, such as ambiguity errors (Appendix A);

f. excessive boat speed;

g. poor GPS data attributed to multipath, satellite changes, or high dilution of precision (DOP);

h. excessive pitch-and-roll or erratic motion of boat and ADCP; and,

i. inadvertent early termination of the transect.

Boat Operation

Average boat speed during each transect normally should be less than or equal to the average water speed. At some sites, it may be necessary to move the boat across the channel using a non-ferrous tag line in order to meet this requirement. Other methods for moving the boat slow enough to be equal to or less than the water speed include the use of push poles, paddles, low-speed trolling motors, or tethered boats, which can be moved slowly across the channel when deployed from a hand-operated cableway or a bridge. In certain conditions, it may not be possible to keep the boat speed less than the water speed. If it is not practical or safe to keep the boat speed less than or equal to the average water speed, additional transects may be measured to obtain a good average discharge. The reason that the boat speed was higher than the average water speed should be documented on the ADCP discharge-measurement field note form. Ongoing research (Oberg and Mueller, 2007a) indicates that the number of transects and the boat speed are not as important as the cumulative time in which data are collected and averaged. A cumulative time for data collection of at least 720 seconds should result in a good mean discharge in steady-flow conditions. When using GPS, keeping the boat speed as low as practical is especially important because errors in the compass readings are additive and increase with boat speed. Rapid course changes should be avoided; the key element in boat operation during the measurement is to do everything slowly and smoothly. Simpson (2002) discusses proper boat operation for ADCP measurements in detail, and his remarks on boat operation should be heeded (Simpson, 2002, p. 122):

> "Be a smooth operator! The BB [broadband]-ADCP discharge-measurement system will give more consistent results if rapid movements and course changes are kept to a minimum. Smooth boat motion is more important than a straight-line course."

Estimating Edge Discharge

Because depths will eventually get too shallow for valid data collection as the ADCP approaches a bank, it is necessary to estimate discharge in the near-shore unmeasured zones using the ADCP discharge-measurement software. To ensure the accuracy of near-shore discharge estimates, the distances from the edge of water to the starting and stopping points of each transect must be measured using a distance-measurement device (such as a laser or optical rangefinder), tagline, or some other accurate measurement device. Placing marker buoys at the start and end points of transects is advantageous for keeping consistent edges. Use of marker buoys enhances the data collection by ensuring more consistent edge estimates and by measuring in approximately the same section for all passes. When measuring in channels with vertical walls at the edges, start and stop points for transects should be no closer to the wall than the depth of water at the wall to prevent acoustic interference from the main beam or side lobes impinging on the wall. For example, if the depth at a vertical wall is 10 ft, transects should start or stop at least 10 ft away from the wall. In order to obtain an accurate mean velocity for estimating the discharge in the near-shore zones, the boat should be kept nearly

stationary from 5 to 10 seconds at the beginning and end of each transect. Accurate edge-discharge estimates also require the ADCP operator to select the correct edge-shape coefficient for the type of edge (sloping or vertical). The edge shapes should be recorded in the ADCP discharge-measurement notes (Oberg and others, 2005).

When using a tethered boat, special methods are required to measure edge distances. Distance marks on the bridge handrail or guardrail may be used to measure edge distances (fig. 11). If the tethered boat is too far away from the bridge to accurately use distance marks for measuring edge distances, laser rangefinders having a compass, an inclinometer, and a "missing-line mode" capability may be used. Missing-line mode calculates a horizontal distance between two points, given a range, heading, and vertical angle measured for each point. Edge distance may be measured by selecting the shore and the transect start or end point while using this mode (Rehmel and others, 2002).

When using a remote-controlled boat at some sites, edge distances may be measured using the same techniques as with tethered boats. At other sites where edge distances cannot be measured using these techniques, it may be necessary to have someone in line with the measurement section to measure the distance from the near-shore edge of water to the starting point and the distance from the ending point to the edge of water on the far shore.

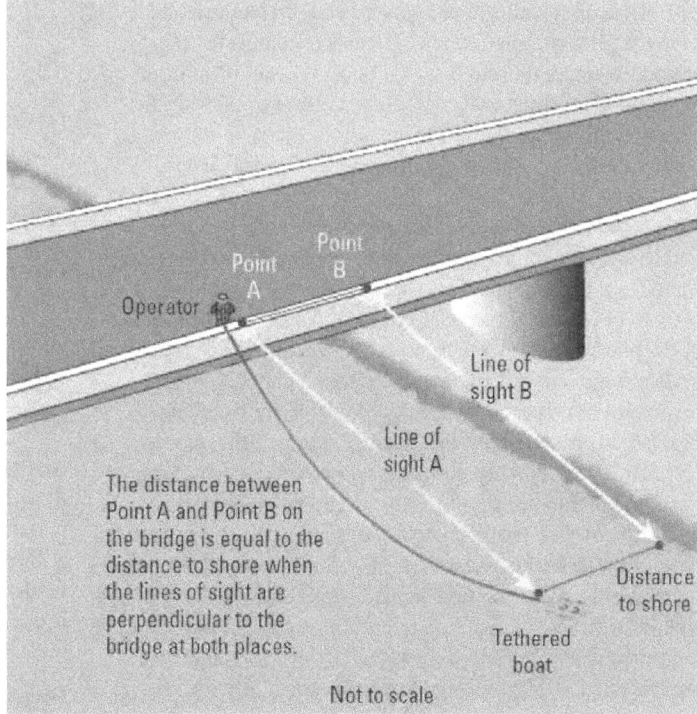

Figure 11. Edge distances needed when using a tethered acoustic Doppler current profiler boat for discharge measurements (modified from Environment Canada, 2004).

Field Notes

All information on an ADCP measurement field note form should be filled out during the course of the measurement. The ADCP operator should note any conditions that potentially could affect the measurement, including estimated wind speed and direction, bi-directional or unusual flow patterns, excessive waves, or passing boats. Use of an ADCP does not negate long-standing, agency guidelines and policies regarding measurement documentation, such as recording reference gage heights before, after, and, if needed, during the discharge measurement. An example of a completed USGS ADCP discharge-measurement field note form is shown in figure 12 (Oberg and others, 2005).

Step-by-Step Procedure

The steps for making a discharge measurement using an ADCP are not complex, but each step must be completed to ensure quality data. To assist the field hydrographer, quick-reference guides that detail the step-by-step procedure for making ADCP discharge measurements, along with other pertinent information, are presented in Appendix E, figures E-3–E-6. These guides can be printed, laminated, and kept with the ADCP for reference.

Post-Measurement Field Procedures

An assessment of the discharge measurement should be made after completion of the transects composing the measurement. A thorough review of all measurement data may not be practical in the field, but a cursory review of the measurement should be made in order to assign a preliminary quality rating to the measurement and to ensure that specific transects do not have critical data-quality problems. If all data were collected at the same measurement section, the transect widths and discharges in the measured (middle) and unmeasured (top, bottom, and edge) sections should be consistent. If transect widths or discharges are not consistent with those of the other transects, the transect data should be scrutinized to determine if a critical data-quality problem occurred (examples of critical data-quality problems are listed in the Discharge-Measurement Procedures section of this report). If a critical data-quality problem is identified, the data from the affected transect should not be used in the computation of discharge. A new transect should be measured, starting from the same side as the discarded transect, if flow conditions have remained steady. If the flow has changed, a new transect series should be collected. A minimum of four transects should be measured if the flow is stable when the new discharge data are collected. A transect should be discarded only if a critical data-quality problem is identified and documented on the field note form. Site-specific conditions, such as turbulence, eddies, reverse flows, surface waves, moving bed, high sediment concentration, and proximity of the instrument to ferrous objects, should be noted under the appropriate sections on the ADCP measurement field note form and used in assigning a quality rating for the measurement (Lipscomb, 1995).

If the discharge measurement was collected at a site with a rating curve, the measured discharge should be plotted on the rating curve for that streamgaging station, and the percentage of difference from the stage-discharge rating should be computed. Rantz and others (1982, p. 346) state: *"If the discharge measurement does not check a defined segment of the rating curve by 5 percent or less, or if the discharge measurement does not check the trend of departures shown by recent measurements, the hydrographer is normally expected to make a second discharge measurement to check his original measurement."* Rantz (1982, p. 346–347) then describes procedures for making check discharge measurements with mechanical current meters. For ADCPs, power off all equipment and begin with step 1c in figure E-3 of Appendix E and proceed through the remainder of the procedures. If practical, choose a new measurement cross section for the check measurement. The measured discharge from the check measurement should then be plotted on the rating curve, and the percentage of difference from the discharge rating should be computed in the field.

Immediately after completion of a measurement, all files, including raw data files, configuration files, instrument test files, compass calibration files, and any electronic measurement forms, should be backed up on nonvolatile media, such as CD-ROMs, flash-memory cards, or USB drives, and stored separately from the field computer. The purpose of this backup is to preserve the data in the event of loss or failure of the field computer.

The ADCP should be dried after use and stored in its protective case for transport. When working in estuaries and other saltwater environments, the ADCP must be rinsed off with freshwater and dried prior to storing for transport. Failure to dry the ADCP may result in corrosion of the ADCP connectors, mounting brackets, and any ADCP accessories stored in the protective case (Oberg and others, 2005).

Office Procedures

Upon returning to the office from field data collection, routine maintenance of equipment should be completed, all data files and notes should be stored properly, data should be reviewed, and measurements should be finalized and archived. Adherence to these procedures will ensure the equipment is ready for the next deployment and that data are reviewed and processed in the most efficient manner.

Preventive Maintenance

The ADCP and associated accessories, such as GPS, vertical depth sounders, and electronic rangefinders, should be inspected upon returning from the field to determine their

Acoustic Profiler Discharge Measurement Notes

Filename Prefix: foxmon_ds1200_

Left Bank: (Sloping) Vertical Other:_____   Right Bank: (Sloping) Vertical Other:_____

Transect No.	Bank	Starting Time	Starting Distance	Ending Distance	Ending Time	Total Discharge	Notes
0	L R						Moving bed test in center of channel
1	L (R)	1249	16	69	1255	1,321	Simultaneous comparison discharge meas.
2	(L) R	1256	69	16	1301	1,358	Upstream of dam
3	L (R)	1301	69				Transect aborted due to debris in river
4	L (R)	1303	69	16	1308	1,327	
5	(L) R	1309	16	69	1315	1,356	
	L R						
	L R						
	L R						
	L R						
	L R						
	L R						
	L R						

Notes: All times are CST. Measurement was made using a temporary rope-and-pulley cableway. Edge distances were measured with laser rangefinder by marking the start and ending positions on the rope and measuring the distance from edge of water to the center of the tethered boat.

U.S. DEPARTMENT OF THE INTERIOR
U.S. Geological Survey
ADCP Discharge Measurement Notes

9-2751   10.2408   Meas. No. 57   Processed by BLL   Checked by KAO

Station Number: 05551540
Station Name: Fox River at Montgomery, IL
Date: July 6, 2004
Party: B.L. Loving, S.E. Anderson
Index Vel.: 1.9   Gage Height: 11.74   Discharge: 1,350
Width: 235   Area / Rated Area: 707
Mean plots: 1.1   From rating: % diff No. 4   ADCP Sync'd to WT   Y at 1207 or N   Shift: 0
Gage Height Change: 0.00 in 0.4 hrs
ADCP Mfr / Model / Frequency: TRDI/Rio Grande/1200   Serial No. 1836   Firmware: WinRiver 10.06   Software: WinRiver 10.06
Boat Motors Used: OceanScience Tethered   GPS Used: Trimble AgGPS   ADCP Depth: 0.27 ft   NavFor Method: Bottom Track
Compass Calib. & Total Error: 24   On-site Model: Previous   Y or N   Diag. Test Errors? Y or N   Moving Boat? Y or (N)
Mean Water Temp: 26.5 T or (C) at 1210   ADCP Water Temp: 25 T or (C) at 1210   Weather / Air Temp: Sunny & clear 83 (F) C   Wind Speed / Dir.: Southerly @ 5-10

Site Conditions
Max Water Depth: 10 ft
Max Water Speed: 2.5 fps
Max Boat Speed: 1 fps
Water Mode: 12
Bottom Mode: 5
Streambed material: Gravel
Salinity: --- ppt at ---
Checkbar found: 22.41
Checkbar changed to: at

Gage Readings

Time	ETG	CRIO	Inside	Outside
1100		11.74	11.70	11.70
1230		11.70		±0.05
1249	(s)	11.70		
1300		11.70		
1315	(f)	11.70		
1400		11.74		11.70 ±0.05

Weighted MGH: 11.74
GH corrections: 11.74
Correct MGH: 11.74

Wading (CGR) Ice boat, upstr, downstr, side bridge: 1,500 (ft) mi. upstr, downstr of gage
Measurement rated: excellent (2%) (Good (5%)) fair (8%), poor (>8%) based on following conditions
Flow: Steady & uniform. Flow at edges appears to be moving in ds direction.
Cross section: Sand and gravel with some mud
Control: Dam is clear of debris
Gage operating: (Y) or N   Record removed: Y or (N)   Filename: telephone telemetry   Intakes/Orifice cleaned/purged: No
Battery voltage: 12.5 V   Tank: Line: Bubble rate:
Bubble-gage psi: Max: Min: Adm:
Extreme-GH indicators: none   Ref. elev.: 12.65   CSG Checked: (Y) or N   HWM elevation: none
HWM on stick: none   -- depth at control --   =   -A-   Raised= 10.48   Raised= none
GH of zero flow = GH: Sheet No. 1 of 1   sheets

Figure 12. Example of completed acoustic Doppler current profiler discharge-measurement field note form (Oberg and others, 2005).

condition. Deployment platforms and mounts also should be inspected. Damage or undue wear to any instrument components, deployment platforms, or mounts should be corrected as soon as possible. The ADCP, all accessories, platforms, mounts, and field computers should be prepared for redeployment and stored in an appropriate location. All batteries should be recharged immediately to facilitate rapid reuse.

Data Storage

All measurement data should be moved from the field computer or field backup media to a permanent storage location for archival and backup. Field computers used to collect ADCP data should have local area network (LAN) capability to facilitate the process of transferring the measurement data to an office server.

Measurement Review Procedures

Discharge measurements should be reviewed in detail by the person who made the measurements as soon as practical after completion of ADCP field measurements. ADCP discharge measurements should be routinely checked by someone other than the person who made the measurement, in accordance to specific agency policies.

Important aspects of reviewing ADCP discharge measurements both in the office and in the field as soon as the data are collected are listed below.

1. The discharge-measurement field note forms should be complete, understandable, and legible.

2. All electronic data files associated with the measurement should be backed up in the field and archived on an office server.

3. The number of transects measured should be appropriate for the flow conditions and satisfy agency policy. Transects should be measured in reciprocal pairs.

4. Configuration files should be checked for errors, appropriateness for the hydrologic conditions, and consistency with field notes. ADCP depth, salinity, edge distances, edge shapes, extrapolation methods, and ADCP configuration parameters listed on the field notes should match those in the configuration file.

5. The temperature measured by the ADCP thermistor should be reasonable for the site and time of year and match the water temperature measured and noted on the field form. Speed-of-sound calculations that are not corrected for temperature can cause velocity-measurement errors and depth errors as great as 7 percent. An error in temperature caused by a faulty ADCP thermistor results in an erroneous calculation

of water density and introduces uncertainty into the speed-of-sound calculations (Simpson, 2002).

6. The salinity of the water at the measurement site should be measured and noted on the field form and entered into the ADCP software for use in the speed-of-sound calculations. If the hydrographer has entered an incorrect salinity value or has forgotten to enter the proper value, depths and velocities will be calculated incorrectly. Errors in excess of 3 percent can be caused by speed-of-sound calculations that are not corrected for salinity (Simpson, 2002).

7. A moving-bed test using proper technique should be performed prior to the discharge measurement, recorded, archived, and noted on the ADCP measurement field note forms. If a moving bed was detected, GPS should be used. If GPS was not used, the measured discharges should be adjusted for the moving bed (Appendix B).

8. The average boat speed for the measurement should not have exceeded the average water speed unless it was impractical or unsafe to do so. The reason for any exceedance should be documented in the field notes or station file. Boat pitch-and-roll should not be excessive. Excessive boat speed or pitch-and-roll may justify downgrading the measurement quality.

9. The measured edge distances recorded on the ADCP measurement field form should match those electronically logged with each transect. The correct edge shape should be selected and 5–10 seconds of data collected at transect stop and start points while the boat is held stationary. If subsectioning was used to correct problems with edges, then the reason for subsectioning should be clearly documented on the field forms. If a vertical wall is present, then the start and end points for the transect should be located such that the distance from the wall is equivalent to or greater than the water depth at the wall.

10. The number of missing or invalid ensembles should not be excessive. (An ensemble is a single profile of the water velocity through the water column consisting of one or the mean of multiple pings.) The number of missing or invalid ensembles that will result in a poor measurement is difficult to establish because the location and clustering of the missing or invalid ensembles is important. If 50 percent of the ensembles were missing or invalid, but every other ensemble was valid, the measurement could still be a good measurement. However, if 10 percent of the ensembles were missing or invalid, but they all occurred in one location where the neighboring valid data would be a poor representation of what was unmeasured, the measurement would be poor. When the missing or invalid ensembles always occur in

the same part of the cross section, and the percentage of flow that is likely unmeasured and, therefore, estimated for missing or invalid ensembles exceeds 5 percent, the measurement quality should be downgraded or the transect determined to be unacceptable.

11. The criteria for invalid depth cells are similar to those for missing or invalid ensembles. Degrading the measurement is not necessary if the distribution of the invalid depth cells is fairly uniform throughout the water column or the measured cross section. However, significant unmeasured portions of the section due to one or more clusters of invalid depth cells can be a reason to downgrade the measurement quality or deem the transect unacceptable.

12. The extrapolation method for the top and bottom discharges should be reviewed. If review of the data shows the need for a different extrapolation method than that chosen for use in the field, the extrapolation method should be corrected and the reasons documented on or attached to the measurement field form. Wind and horizontally stratified density currents are common causes for profiles that poorly fit the one-sixth power-law extrapolation method. In these situations, it is usually necessary to use different extrapolation techniques for the top and bottom areas and(or) to limit the portion of the profile used for the selected method.

    a. RiverSurveyor software allows the user to determine which portions of the profile are used to estimate the top and bottom unmeasured areas using either a constant or one-sixth power-law method. For irregular profiles, adjusting the number of points used for the top or bottom extrapolation can provide a better estimate of the unmeasured discharge.

    b. WinRiver software provides different options for the top and bottom extrapolation. The three-point slope method for top extrapolation uses the top three depth cells to estimate a slope, and this slope is then applied from the top depth cell to the water surface; however, a constant value or slope of zero is assumed if less than six depth cells are present in the profile. The velocity must go to zero at the streambed, and this is often referred to as a "no-slip" condition in fluid mechanics. The no-slip method for bottom extrapolation uses the depth cells present in the lower 20 percent of the depth to determine a power fit forcing it to zero at the bed. In the absence of any depth cells in the lower 20 percent, the no-slip method uses the last single valid depth cell and forces the power fit through it to be zero at the bed. It is impor-

tant to also note that the bottom extrapolation method chosen by the user determines how missing depth cells in the center of the profile are estimated. If the power fit is used at the bottom, the velocities for the invalid depth cells are interpolated using the power fit through all of the data. If the bottom extrapolation method is set to no slip, a linear interpolation method is used to fill in invalid data.

13. Measurement computations, including mean discharge and measurement gage height, must be correct.

Problems identified during the review process should be viewed as an opportunity for improving future measurements. If the measurement section had undesirable characteristics (undesirable measurement section characteristics are described in the Site Selection section of this report), future measurements should be made at a more appropriate measurement section. If boat operation technique problems are identified, these problems should be discussed with the boat operator so they are not repeated during future measurements. If the ADCP and the data-collection software could have been more accurately configured, this problem also should be discussed with the field crew. A step-by-step measurement review process is summarized in Appendix F – Measurement Review Procedures.

Data-Quality Indicators

Data-quality indicators for ADCP discharge measurements are not only valuable in the office during measurement processing but also are helpful in the field for selecting measurement sites and identifying problems and making necessary adjustments while data are being collected. Possible data-quality indicators follow.

- Numerous missing ensembles indicate communication problems between the ADCP and the computer. The computer may not be able to keep up with the incoming data or may have gone to sleep and missed incoming data. Power to the ADCP may have been interrupted or the communications cable may not have been attached securely. Missing data cannot be recovered.

- Numerous invalid ensembles indicate that the ADCP frequently was unable to measure the velocities in a portion of the cross section (figs. 13 and 14). For an invalid ensemble, the software receives all of the data from the ADCP, but the data do not meet the criteria for a valid velocity measurement. Invalid ensembles can be caused by (a) invalid bottom tracking, which would provide no boat reference from which to compute the velocity, (b) decorrelation of the acoustic pulse (from turbulence, high shear, submerged debris, or fish), which would not permit an accurate measurement of the Doppler shift, (c) low backscatter, which

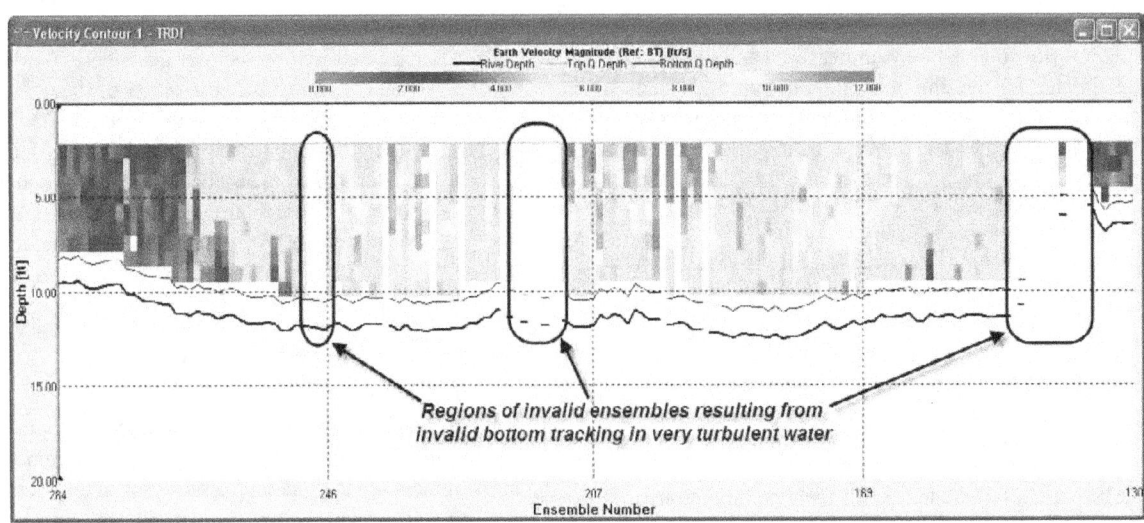

Figure 13. Screen capture from Teledyne RD Instruments WinRiver software illustrating numerous invalid ensembles collected in the Pigeon River at Canton, North Carolina, as a result of invalid bottom tracking.

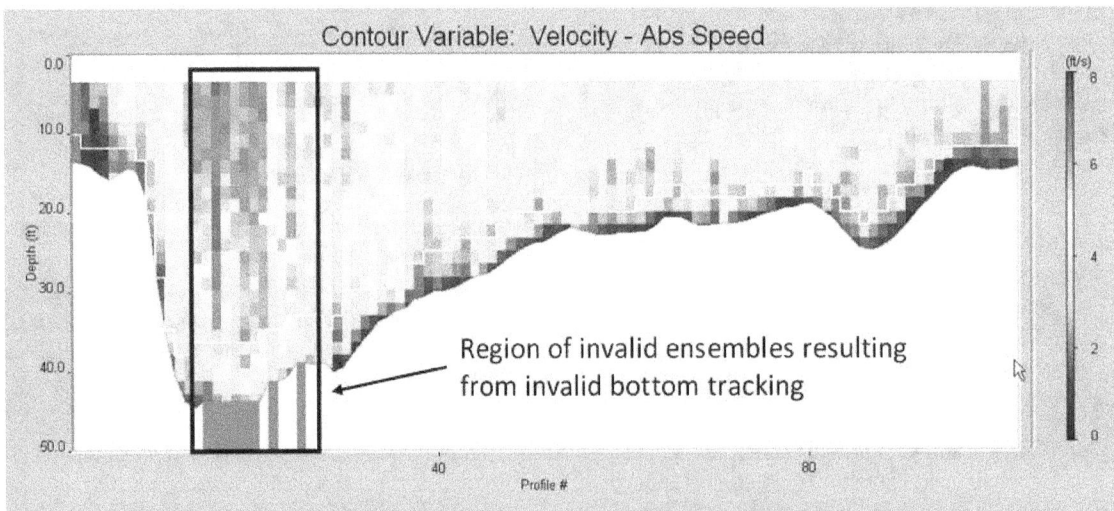

Figure 14. Screen capture from Sontek/YSI RiverSurveyor software illustrating numerous invalid ensembles in the thalweg of the Mississippi River at Chester, Illinois, as a result of invalid bottom tracking.

results in an insufficient amount of acoustic energy reflected back to the transducer to allow the ADCP to measure the Doppler shift, (d) the blocking of acoustic pulses by air entrainment, or (e) user-specified data-quality criteria.

- Ambiguity errors in velocity measurements will show up as velocity spikes when compared with the neighboring valid velocity measurements (fig. 15). Typically, this is only a problem for broadband instruments and can be filtered out in post processing.

- Beam intensities should be consistent throughout the water column down to the streambed, at which point a relatively large intensity should be evident because the bottom is a much better reflector of acoustic energy than scatterers in the water. Anomalies in beam intensities through the water column can indicate interference from side walls, fish, trees, or other submerged debris that can degrade the quality of the velocity measurement (figs. 16 and 17).

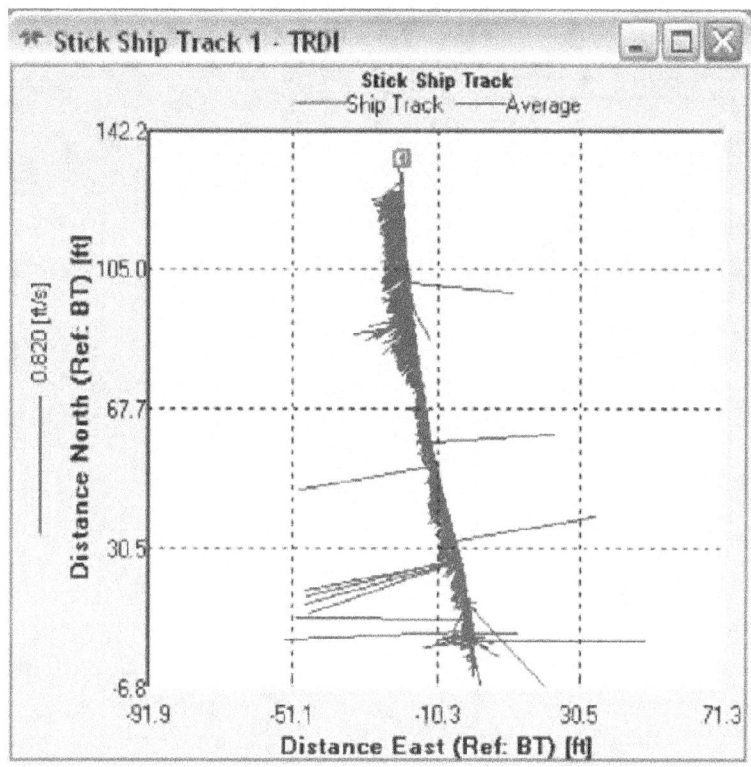

Figure 15. Screen capture from Teledyne RD Instruments WinRiver software illustrating erroneous velocity measurements caused by ambiguity errors.

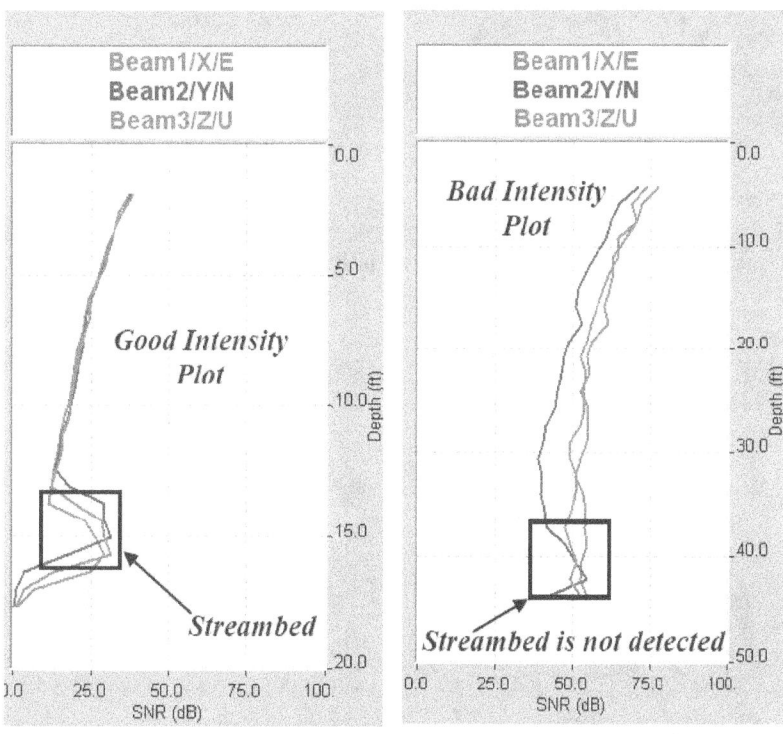

Figure 16. Screen capture from Sontek/YSI RiverSurveyor software illustrating an example of good and bad beam intensity data relative to detection of the streambed.

• Irregular or erratic boat motion creates rapid horizontal accelerations of the ADCP, which leads to noisy boat velocity measurements and in turn can degrade the accuracy of the water-velocity measurement. The most common cause of irregular or erratic boat motion is shifting the boat in and out of gear frequently (fig. 18). The consistency of boat speed throughout a cross section is most evident in the time series plot of water and boat speed (figs. 18 and 19).

• Rapid variation in the pitch-and-roll time series plots indicate that the ADCP is measuring in rough, turbulent conditions or that there is significant movement of people or equipment in the boat deploying the ADCP. Inconsistent pitch-and-roll throughout an ADCP measurement can lead to decorrelation of the acoustic signal and result in numerous invalid ensembles (unmeasured sections of the channel).

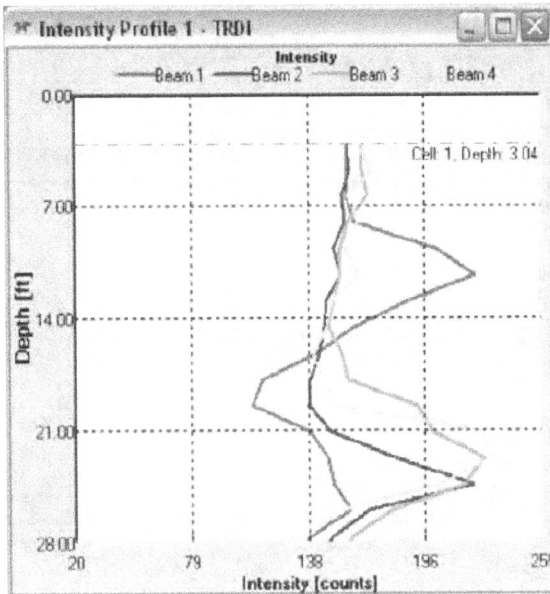

Figure 17.   Screen capture from Teledyne RD Instruments WinRiver software illustrating an anomaly in beam intensities caused by interference in beam 1 from a side wall.

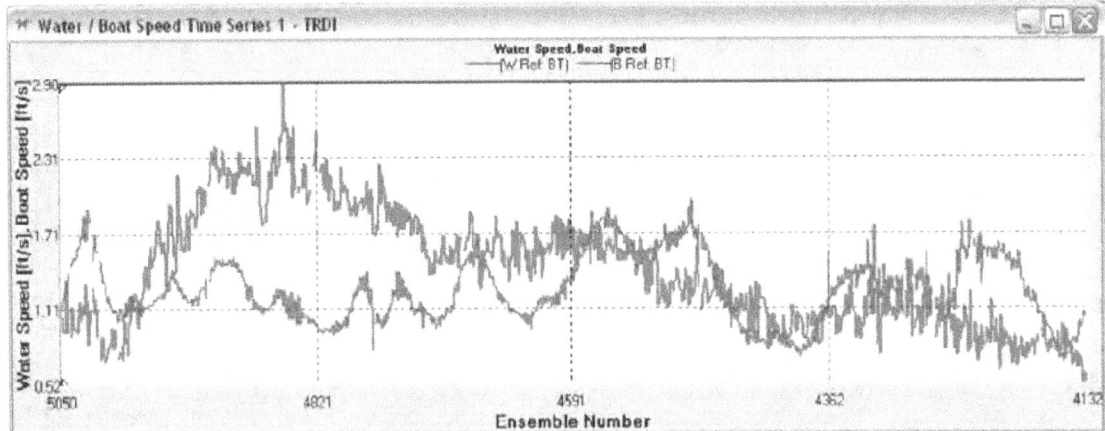

Figure 18.   Screen capture from Teledyne RD Instruments WinRiver software illustrating highly variable boat speeds resulting from shifting the motor in and out of gear while measuring along the transect (red line–water speed; green line–boat speed).

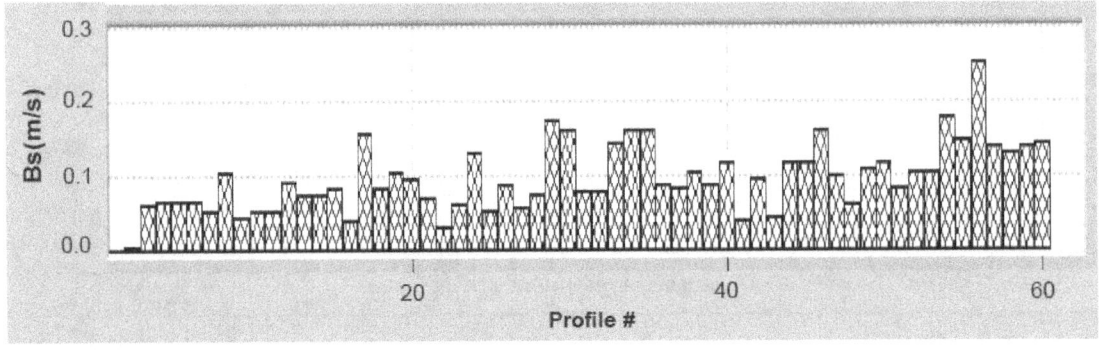

Figure 19.   Screen capture from Sontek/YSI RiverSurveyor software illustrating variation in boat speed while measuring along a transect.

• Irregularities in bottom-tracking depth measurements will lead to spikes in the streambed profile (fig. 20), which can influence the accuracy of measured discharge by biasing the measured cross-sectional area. The depth used for discharge computations is an average of the depths measured for each of the ADCP beams. If one or more of the beams strikes an object such as a fish, a submerged tree, or debris, before reflecting off the streambed, or if the ADCP processes a multiple-return of the bottom acoustic pulse, the recorded depth for that section and therefore the cross-sectional area will be biased. Post-processing data-screening functions in the software can smooth these spikes in bottom depth.

• When using the GGA output string from GPS as a boat-speed reference, the quality of the differentially corrected global positioning system (DGPS) signal influences the measurement of ADCP movement, which directly affects the quality of the water-velocity measurement by the ADCP (fig. 21). The integration of a DGPS to track the movement of the ADCP can be used to avoid the systematic bias associated with a moving bed. DGPS systems, however, cannot always provide consistently accurate positions because of multipath errors, changes in viewable satellites, and satellite signal reception problems on waterways with dense tree canopy along the banks, in deep valleys or canyons, and near bridges. Hydrographers should monitor the DGPS quality-assurance tabular summaries provided in ADCP software to assure that the DGPS signal is not affected by these errors during data collection.

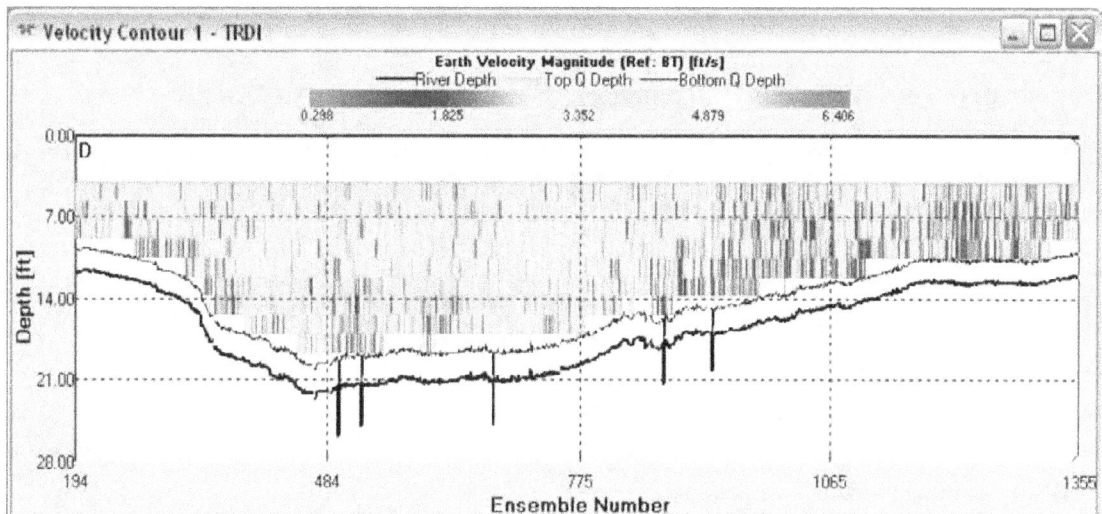

Figure 20. Screen capture from Teledyne RD Instruments WinRiver software illustrating spikes in the streambed profile.

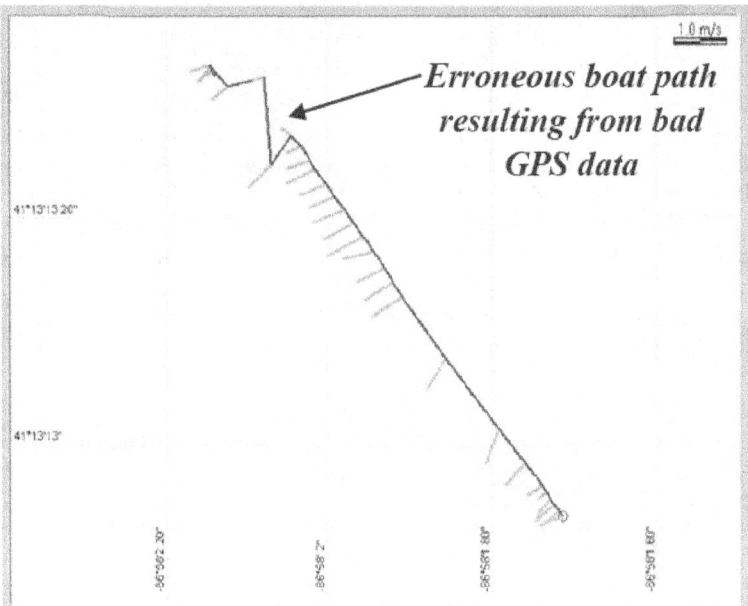

Figure 21. Screen capture from Sontek/YSI RiverSurveyor software illustrating the effects of poor GPS data on the measurement of boat movement.

Commonly Observed Measurement Problems

In-depth reviews of acoustic programs at USGS offices indicate that the most common problems related to ADCP data quality include the following:

1. No moving-bed test conducted or documented;

2. Discharge measurements not adjusted or down-graded correctly when a moving-bed condition is present;

3. Selection of a cross section with site conditions that are not appropriate for the ADCP;

4. The instrument mode that is most applicable to site conditions is not used;

5. Excessive boat velocity (appreciably greater than the mean water velocity);

6. Edge distances estimated, not measured;

7. Poor data-archival procedures;

8. Incorrect extrapolation methods for profile shape;

9. No ADCP diagnostic test;

10. ADCP time not properly set;

11. Poor field notes; and

12. ADCP depth not measured or incorrectly assigned.

ADCP users should pay special attention to these problems when planning ADCP data-collection efforts, collecting data, processing and archiving data, and developing office quality-assurance plans (Oberg and others, 2005).

Selected References

Callede, J., Kosuth, P., Guyot, J.L., and Guimaraes, V.S., 2000, Discharge determination by acoustic Doppler current profilers (ADCP): A moving bottom error correction method and its application on the River Amazon at Obidos: Hydrological Sciences-Journal-des Sciences Hydrologiques, v. 45, no. 6, p. 911–924.

Chen, Cheng-Lung, 1989, Power law of flow resistance in open channels—Manning's formula revisited: Proceedings of the International Conference on Channel Flow and Catchment Runoff, Centennial of Manning's Formula and Kuichling's Rational Formula, May 22–26, 1989, Charlottesville, VA, v. 8, p. 17–48.

Christensen, J.L., and Herrick, L.E., 1982, Mississippi River test, volume 1. Final report DCP4400/300, prepared for the U.S. Geological Survey by AMETEK/Straza Division, El Cajon, California, under contract No. 14–08–001–19003, p. A5–A10.

Doppler, J.C., 1842, Über das farbige Licht der Doppelsterne und einiger anderer Gestirne des Himmels: Abn.königl. böhm.Ges.Wiss, v. 2, p. 465–482.

Environment Canada, 2004, Procedures for conducting ADCP discharge measurements: Water Survey of Canada, Hydrometric Operations Division, SOP001–2004.

Federal Aviation Administration, 2006, WAAS and its relation to enabled hand-held GPS receivers, accessed September 22, 2008, at *http://gpsinformation.net/exe/waas.html*

Fulford, J.M., and Sauer, V.B., 1986, Comparison of velocity interpolation methods for computing open-channel discharge *in* Subitsky, S.Y., ed., Selected papers in the hydrologic sciences: U.S. Geological Survey Water-Supply Paper 2290, 154 p.

García, C.M., Oberg, K., and García, M.H., 2007, ADCP measurements of gravity currents in the Chicago River, Illinois: Journal of Hydraulic Engineering, v. 133, no. 12, p. 1356–1366.

Huang, Hening, 2000, Principle and accuracy analysis of river discharge measurement using SonTek RiverSurveyor System: San Diego, CA, SonTek/YSI, 11 p.

Lipscomb, S.W., 1995, Quality assurance plan for discharge measurements using broad-band acoustic Doppler profilers: U.S. Geological Survey Open-File Report 95–701, 7 p.

Mathworks, Inc., 2005, Matlab—The language of technical computing: Natick, MA, Mathworks, Inc., v. 7.1.

Morlock, S.E., Nguyen, H.T., and Ross, J.H., 2002, Feasibility of acoustic Doppler velocity meters for the production of discharge records from the U.S. Geological Survey streamflow-gaging stations: U.S. Geological Survey Water-Resources Investigations Report 01–4157, 59 p.

Mueller, D.S., 2002, Use of acoustic Doppler instruments for measuring discharge in streams with appreciable sediment transport, *in* Conference of Hydraulic Measurements and Experimental Methods, Estes Park, CO, 2002, Proceedings: Environmental and Water Resources Institute of the American Society of Civil Engineers.

Mueller, D.S., Abad, J.D., García, C.M., Gartner, J.W., García, M.H., and Oberg, K.A., 2007, Errors in acoustic Doppler profiler velocity measurements caused by flow disturbance: Journal of Hydraulic Engineering, v. 133, no. 12, p. 1411–1420.

Mueller, D.S., and Wagner, C.R., 2006, Application of the loop method for correcting acoustic Doppler current profiler discharge measurements biased by sediment transport: U.S. Geological Survey Scientific Investigations Report 2006–5079, 26 p.

National Marine Electronics Association, 2002, NMEA 0183 Standard for interfacing marine electronic devices, version 3.01: National Marine Electronics Association.

Oberg, K.A., 2002, In search of easy-to-use methods for calibrating ADCPs for velocity and discharge methods, *in* Wahl, T.L., Pugh, C.A., Oberg, K.A., and Vermeyen, T.B., eds., 2002, Hydraulic measurements and experimental methods 2002: Proceedings, Conference of Environmental and Water Resources Institute of the American Society of Civil Engineers, July 28–August 1, 2002, Estes Park, CO.

Oberg, K.A., Morlock, S.E., and Caldwell, W.S., 2005, Quality-assurance plan for discharge measurements using acoustic Doppler current profilers: U.S. Geological Survey Scientific Investigations Report 2005–5183, 35 p.

Oberg, K.A., and Mueller, D.S., 2007a, Analysis of exposure time on streamflow measurements made with acoustic Doppler current profilers, *in* Proceedings of the Hydraulic Measurements and Experimental Methods 2007, Reston, VA, American Society of Civil Engineers.

Oberg, K.A., and Mueller, D.S., 2007b, Validation of streamflow measurements made with acoustic Doppler current profilers: Journal of Hydraulic Engineering, v. 133, no. 12, p. 1421–1432.

Ott, Lyman, 1988, An introduction to statistical methods and data analysis: Boston, PWS-KENT Publishing Company, 945 p.

Rantz, S.E., and others, 1982, Measurement and computation of streamflow, volume 1, Measurement of discharge: U.S. Geological Survey Water-Supply Paper 2175, 631 p.

Rehmel, M.S., Stewart, J.A., and Morlock, S.E., 2002, Unmanned, tethered acoustic Doppler current profiler platforms for measuring streamflow: U.S. Geological Survey Open-File Report 03-237, 15 p., accessed September 22, 2008, at *http://hydroacoustics.usgs.gov/publications/rehmel_teth_dop.pdf*

Ruhl, C.A., and Simpson, M.R., 2005, Computation of discharge using the index-velocity method in tidally affected areas: U.S. Geological Survey Scientific Investigations Report 2005–5004, 41 p.

Schlichting, H., 1979, Boundary layer theory (7th ed.): New York, McGraw-Hill.

Simpson, M.R., 2002, Discharge measurements using a broadband acoustic Doppler current profiler: U.S. Geological Survey Open-File Report 01–01, 123 p., accessed June 22, 2007, at *http://pubs.usgs.gov/of/2001/ofr0101/*

Simpson, M.R., and Oltmann, R.N., 1993, Discharge measurement using an acoustic Doppler current profiler: U.S. Geological Survey Water-Supply Paper 2395, 34 p.

SonTek/YSI, 2000, Acoustic Doppler profiler principles of operation: San Diego, CA, SonTek/YSI, 28 p.

SonTek/YSI, 2007, RiverSurveyor system manual, software version 4.60: San Diego, CA, SonTek/YSI, 194 p.

Teledyne RD Instruments, 1996, Principals of operation—A practical primer for broadband acoustic Doppler current profilers (2d ed.): San Diego, CA, Teledyne RD Instruments, 51 p.

Teledyne RD Instruments, 2003, WinRiver user's guide— USGS version: San Diego, CA, Teledyne RD Instruments, P/N 957–6096–00, 156 p.

Teledyne RD Instruments, 2006, WinRiver II quick start guide: San Diego, CA, Teledyne RD Instruments, P/N 957–6230–00, 37 p.

Teledyne RD Instruments, 2007, WinRiver II user's guide: San Diego, Teledyne RD Instruments, CA, P/N 957–6231–00, 166 p.

Teledyne RD Instruments, 2008, Workhorse Rio Grande ADCP, accessed December 15, 2008, at *http://www.rdinstruments.com/datasheets/rio_grande_ds_lr.pdf*

Urick, R.J., 1983, Principles of underwater sound (3d ed.): New York, McGraw-Hill, 423 p.

U.S. Army Corps of Engineers, 2002, Engineering design manual on hydrographic surveying (EM 11103–2–1003), accessed August 24, 2006, at *http://www.usace.army.mil/inet/usace-docs/eng-manuals/em1110-2-1003/toc.htm*

U.S. Geological Survey, 2002a, Configuration of acoustic profilers (RD Instruments) for measurement of streamflow: U.S. Geological Survey, Office of Surface Water Technical Memorandum 2002.01, accessed August 13, 2008, at *http://hydroacoustics.usgs.gov/memos/OSW2002-01.pdf*

U.S. Geological Survey, 2002b, Policy and technical guidance on discharge measurements using acoustic Doppler current profilers: U.S. Geological Survey, Office of Surface Water Technical Memorandum 2002.02, accessed August 13, 2008, at *http://hydroacoustics.usgs.gov/memos/OSW2002-02.pdf*

U.S. Geological Survey, 2003, Release of WinRiver software version 10.05 for computing streamflow from acoustic profiler data: U.S. Geological Survey, Office of Surface Water Technical Memorandum 2003.04, accessed July 21, 2008, at *http://hydroacoustics.usgs.gov/memos/OSW2003-04.pdf*

U.S. Geological Survey, 2005a, Guidelines for archiving electronic discharge measurement data: U.S. Geological Survey, Office of Surface Water Technical Memorandum 2005.08, accessed August 13, 2008, at *http://hydroacoustics.usgs.gov/memos/OSW2005-08.pdf*

U.S. Geological Survey, 2005b, Guidance on the use of RD Instruments StreamPro acoustic Doppler profiler: U.S. Geological Survey, Office of Surface Water Technical Memorandum 2005.05, accessed August 13, 2008, at *http://hydroacoustics.usgs.gov/memos/OSW2005-05.pdf*

Appendix A – Basic ADCP Operational Concepts

Descriptions of how an ADCP operates and measures water and boat velocity can be found in reports produced by ADCP manufacturers (SonTek/YSI, 2000; Teledyne RD Instruments, 1996), by the USGS (Simpson, 2002), and by others not referenced herein. This appendix is not intended to be a comprehensive coverage of information that can be found in other documents, but rather a concise overview of key operation concepts that will help the user understand the capabilities and limitations of ADCPs.

General

The ADCP transmits acoustic energy at a known frequency and measures the change in frequency of the acoustic energy reflected back (backscattered) from particles in the water column. The velocity of the water along the acoustic path can be computed from equation A1.

$$V = \frac{CF_D}{2F_S},\qquad (A1)$$

where

V	is the velocity of the water parallel to the acoustic path;
C	is the speed of sound in the water;
F_D	is the difference in frequency due to the Doppler shift (F_B–F_S);
F_S	is the frequency of the transmitted acoustic energy; and
F_B	is the frequency of the backscattered acoustic energy.

The success of the Doppler approach to measuring water velocity rests on the assumption that a sufficient amount of material is in the water column to reflect enough acoustic energy to allow the measurement of the Doppler shift and that the material is traveling at the same velocity as the water. If an object in the water column is large relative to the suspended material and has a velocity that is not dependent on the water velocity (such as fish or woody debris lodged on bottom) the acoustic energy reflected from this object will not have a Doppler shift that reflects the water velocity and will likely corrupt the velocity profile at and below the depth of the object. Most instruments contain algorithms to detect and filter out data with these types of errors.

Measuring Velocity

Currently (2008) the two techniques commonly used in an ADCP to measure the Doppler shift can be classified as narrowband or broadband. Each approach has its advantages and disadvantages. A detailed explanation of each technique is beyond the scope of this report, and some of the specifics of their implementation into commercial instruments are considered proprietary. The basics of the two approaches, however, are described using an analogy developed by Joel Gast (Teledyne RD Instruments, oral commun., 1992) and published by Simpson (2002).

Narrowband

Consider a freeway at night with traffic moving at a steady rate of speed. A camera has been placed near the freeway, and posts have been installed at set distances within the camera's field of view. A strobe light is actuated and, while the freeway is illuminated by the single-strobe pulse, the camera takes two high-speed photographs. An investigator examines the photographic negatives and lines up the images of the cars on the two photographic negatives. The investigator then determines the distance traveled by the cars by measuring the apparent shift in position of the reference posts (fig. A-1). The speed of the cars can be calculated by dividing distance between the posts by the lag time between the two photographs. If the strobe flashes represent acoustic pulses, the cars represent reflective particles in the water column, and the photographic negatives represent the received reflected signals, this scenario becomes roughly analogous to the workings of a narrowband ADCP system.

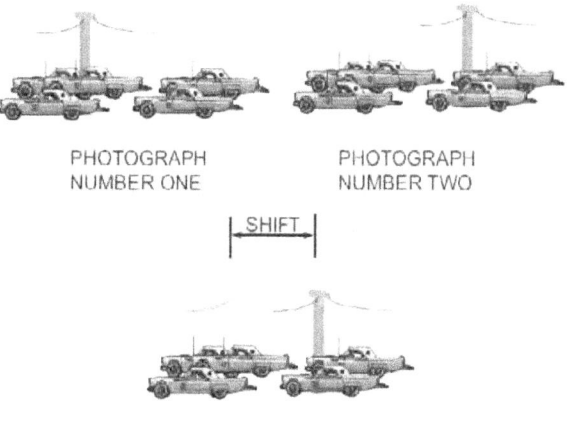

PHOTOGRAPH NUMBER ONE

PHOTOGRAPH NUMBER TWO

SHIFT

COMPOSITE PHOTOGRAPH

Figure A-1. Freeway strobe-light system used to measure vehicle speed (from Simpson and Oltmann, 1993).

The drawback to this approach is that the strobe (pulse length) dissipates very quickly and the two photographs must be taken while the same cars are still illuminated by the strobe. This means that time lag between negatives (received signals) is very short and the distance traveled by the cars (reflectors) is very short; therefore, the car speeds cannot be measured precisely. Because of these limitations, velocity measurements made using the narrowband technology are "noisy," which means they have a relatively high random error. The signal is sampled twice during the reception of the reflected signal. Using an autocorrelation technique, the Doppler shift is then calculated. In the narrowband ADCP, the pulse length depends on the time lag between samples, which is a function of depth-cell size. A filter scheme that looks at the whole returned signal is used to resolve ambiguity.

The high random errors of the narrowband system are overcome by averaging. The standard deviation of random noise about a true mean is reduced by the square root of the number of measurements used to computed the mean (Ott, 1988). The narrowband ping is simple and can be processed quickly. Current (2008) narrowband systems can ping at rates up to 20 Hz. By pinging fast and averaging the velocity measurement of each ping, a narrowband ADCP can measure water velocity with random noise standard deviations of between 0.4 ft/s and 1 ft/s for 1-second averaging and between 0.16 ft/s and 0.43 ft/s for commonly used 5-second averages.

Broadband

Using the same freeway analogy as before, the investigator installs another camera a distance of 10 or more car lengths (parallel to the freeway) from the first camera. He can now actuate a strobe, take a picture with the first camera, wait a short time, actuate another strobe, and take a picture with the second camera. If the strobes are timed correctly, the cars will travel from the field of view of the first camera into the field of view of the second camera during the time between photographs. The investigator synchronizes the positions of the cars on the two negatives and finds that there is a much longer lag time (time between each strobe versus the time between two photographs taken during the same strobe) and that the cars traveled a longer distance. This longer distance of travel and lag time allows the investigator to calculate the speed of the cars with much greater precision than with the single-strobe system. However, the distance between the cameras and the time between each strobe must be chosen carefully. If the investigator waits too long between strobes, random movement between the cars (passing, slowing down, speeding up, and so forth) will render the two negatives "unmatchable" (uncorrelated). Transmitting a pair of pulses (strobes) into the water allows for much longer lag times (therefore, more precision) than the narrowband system, but some disadvantages are associated with this technique.

One of the most significant disadvantages is self noise. Self noise again can be described using the freeway analogy. Suppose that, because of limitations in photographic technology, the freeway cameras have no shutters. Because the investigator must leave the camera shutters open, both cameras will "see" the traffic illuminated by the two strobes. However, only 50 percent of the "scenery" will be usable to both cameras for correlation purposes. For example, the film in camera one is exposed once during the first strobe. The cars then travel out of the field of view of camera one and into the

field of view of camera two. However, the film in camera two already has been exposed by the flash of the first strobe and, thus, any cars photographed have left the field of vision. When the second strobe flashes, the film in both cameras is again exposed (double exposed) and the cars that were first photographed by camera one are now photographed by camera two. Because the film has been double exposed, only 50 percent of the scenery in each exposure contains cars that are common to both cameras. As in the film of the freeway cameras, the reflected wave front from the first broadband ADCP (BB-ADCP) pulse-pair is again "exposed" by the incident wave front of the second pulse and, therefore, is subject to the same "double exposure." The increased noise due to this 50-percent correlation is reduced by data averaging (very narrow pulses can be used, and, therefore, large amounts of data can be collected and averaged). Without a technique called phase coding and a high signal-processing rate, BB-ADCP velocity measurements would be less precise (because of self noise) than measurements made by the narrowband ADCP system.

The broadband approach makes use of very narrow (short) pulses. If these pulses were so narrow that 100 of them could be placed into the space occupied by the original long pulses, the measurement precision would be increased by the square root of the number of samples (in the case of 100 samples, by a factor of 10). With this increased precision (even with the 50-percent level of self noise), the broadband-ADCP measurement precision surpasses the precision of the narrowband ADCP by almost one order of magnitude. However, energy loss caused by the narrow pulses is so great that it renders the system nearly unusable. To overcome this energy loss, the manufacturer developed a design innovation that incorporates most of the advantages of wide and narrow pulses. A wide pulse is transmitted (therefore, delivering more energy into the water than a narrow pulse), but is logically split into many small segments called code elements, each having a phase shift of either 0 or 180 degrees (fig. A-2). The coding order of these phase shifts is pseudorandom (behaves numerically like a random sequence). The consequence of transmitting this phase-coded pulse-pair series into the water is that even though the pulses are long, the signal processor still

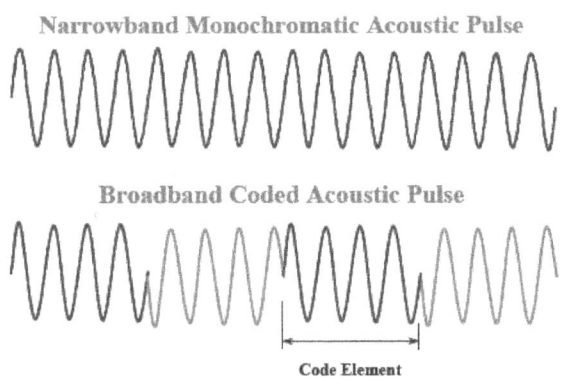

Figure A-2.   Acoustic pulses for narrowband and broadband ADCP technologies.

must wait the full lag period before achieving an auto-correlation peak of significant amplitude. In other words, because of the phase coding, it is difficult for the autocorrelation algorithm to realize a peak at the wrong interval. The objective of the manufacturer is to achieve decorrelation of adjacent pulse pairs and, therefore, a greater effective N (number of samples used for data averaging). This pseudorandom code allows many independent samples to be collected from a single ping. The principal reason this technique was named "broadband" is that the bandwidth of the ADCP was increased to accommodate the signal processing of a narrow-pulse pair (coded or not).

Another significant disadvantage for using the increased lag spacing available with the broadband ADCPs is velocity ambiguity. For this analogy, assume that the distance the cars move in the synchronized photographs is measured using an electronic wheel, rather than a flat ruler. The process is to push a button to zero the wheel, move the wheel along the distance to be measured, and then push another button for the number of degrees of rotation of the wheel to be reported. By knowing the circumference of the wheel and the amount of rotation, we can compute the distance measured. This electronic wheel does not count the number of rotations; it only measures the rotation of the wheel from 0 to 360 degrees. Because the cars could move in either direction, the measurement is further limited to only half of the wheel, so that half of the wheel measures a positive direction and half measures a negative direction, say from 0 to 180 degrees for the positive direction and from 360 to 180 degrees (0 to –180) for the negative direction. This technique works well provided that the relation between the car's speed and lag time between the photographs does not result in more movement than can be measured using 180 degrees of rotation of the wheel. If the lag is too long and(or) the cars are traveling too fast, the distance between the photographs could result in the distance being longer than can be measured by 180 degrees of rotation of the wheel. For example, the distance to be measured requires 190 degrees of rotation, and based on the established reference, 190 degrees would be interpreted as –170 degrees, and the wrong velocity in the wrong direction would be computed. When the rotation exceeds 180 degrees, the velocity measured is ambiguous. It is unkown if the car traveled a distance equal to 190 degrees, –170 degrees, or 550 degrees. The "ambiguity velocity" is the velocity the cars must achieve before this confusing circumstance happens. If the strobe flashes are temporally close, the ambiguity velocity is high (faster than the cars normally travel), but measurement precision is lower because the cars have traveled a shorter distance between strobe flashes. If the time between strobe flashes (lag) is lengthened to improve the measurement precision, the ambiguity velocity becomes lower and limits the maximum velocity that can be measured. How the ambiguity velocity is set or determined in the ADCP is dependent on the water or bottom mode used. Different modes resolve ambiguity differently (see Appendix C for a discussion of water modes).

Computing Velocity in Orthogonal Coordinates

The Doppler shift is directional and can only measure the velocity of motion parallel to the acoustic path (radial velocity; fig. A-3). The goal of most engineering applications is to have three-dimensional velocity components measured in an orthogonal coordinate system. To measure a three-dimensional velocity profile requires a minimum of three acoustic beams

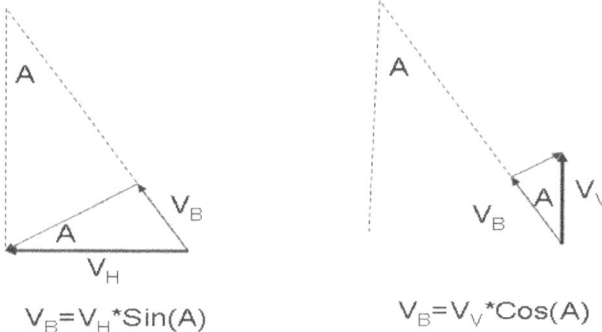

$$V_B = V_H * Sin(A) \qquad\qquad V_B = V_V * Cos(A)$$

Figure A-3. The horizontal (V_H) and vertical velocity (V_V) components computed from the velocity measured parallel to the acoustic path (V_B).

pointed in known directions. The three beams provide three velocity vectors that can then be resolved into three orthogonal velocity vectors. The beams typically are equally distributed around a circular instrument (120-degree spacing) and tilted at a known angle, θ, to the vertical. The equations for a typical three-beam system with beam 1 pointed downstream are:

$$V_y = (2B_1 - B_2 - B_3) / (3\sin\theta), \qquad (A2)$$

$$V_x = (B_3 - B_2) / (\sqrt{3}\sin\theta), \text{ and} \qquad (A3)$$

$$V_z = (B_1 + B_2 + B_3) / (3\cos\theta), \qquad (A4)$$

where

V_y is the streamwise velocity assuming beam 1 is pointed downstream;

V_x is the cross-stream velocity assuming beam 1 is pointed downstream;

V_z is the vertical velocity;

B_1 is the radial velocity measured in beam 1;

B_2 is the radial velocity measured in beam 2;

B_3 is the radial velocity measured in beam 3; and

θ is the tilt angle of the beams referenced to vertical.

The use of four beams in a Janus configuration also is common in ADCPs. The Janus configuration provides some reduction in errors caused by pitch-and-roll and random instrument noise for the orthogonal velocity components. The following equations are for a typical four-beam system with beam 3 pointed forward:

$$V_y = (B_4 - B_3) / (2\sin\theta), \quad\quad (A5)$$

$$V_x = (B_1 - B_2) / (2\sin\theta), \text{ and} \quad\quad (A6)$$

$$V_z = (B_1 + B_2 + B_3 + B_4) / (4\cos\theta), \quad\quad (A7)$$

where

V_y is the streamwise velocity assuming beam 3 is pointed upstream;
V_x is the cross-stream velocity assuming beam 3 is pointed upstream;
V_z is the vertical velocity;
B_1 is the radial velocity measured in beam 1;
B_2 is the radial velocity measured in beam 2;
B_3 is the radial velocity measured in beam 3;
B_4 is the radial velocity measured in beam 4; and
θ is the tilt angle of the beams referenced to vertical.

The assumption in applying equations A2–A7 is that the flow measured in each beam is the same (homogeneous). If the flow measured by one beam is different from any of the other beams (one beam is measuring in a vortex and the others are in the free stream), the basic assumption of these equations is violated, and an incorrect velocity will be computed. There-fore, the velocities measured by an ADCP are less certain in flow conditions with high and rapid spatial variations.

Four-beam systems have a redundant beam that can be used to compute an error velocity. The error velocity is computed as the difference between the vertical velocity computed by opposing beam pairs (3 and 4 versus 1 and 2). Adding the opposing pairs cancels the horizontal velocity and leaves only the vertical component.

$$V_e^u = \frac{B_1 + B_2}{2\cos\theta} - \frac{B_3 + B_4}{2\cos\theta}$$
$$= (B_1 + B_2 - B_3 - B_4) / (2\cos\theta), \quad\quad (A8)$$

where

V_e^u is the unscaled error velocity.

TRDI scales the error velocity so that the standard deviation for the error velocity is equal to the standard deviation of the horizontal velocity. The equation used to compute the scaled error velocity is

$$V_{error} = \left[(B_1 + B_2 - B_3 - B_4) / (2\cos\theta) \right] \left[1 / (\sqrt{2}\tan\theta) \right] \quad (A9)$$
$$= (B_1 + B_2 - B_3 - B_4) / 2\sqrt{2}\sin\theta),$$

where

V_{error} is the scaled error velocity computed by TRDI ADCPs.

Measuring a Velocity Profile

The ADCP profiling capability is accomplished by time gating (and sampling) the received acoustic signal at time intervals as the acoustic-beam wave vertically traverses the water column. The transmitted acoustic pulse travels at the speed of sound through the water and is reflected back toward the transducer. Acoustic energy reflected from particles deeper in the water column takes a longer time to return to the transducer than does acoustic energy reflected from particles nearer the transducer. By measuring the two-way travel time from transmission of the energy to when the energy is received, the distance from the transducer from which the energy is reflected can be computed. The ADCP transmits a ping along each acoustic beam and then time gates the reception of the returned echo on each beam into depth cells at specified ranges. Speed and direction are then calculated (using a center-weighted mean of the velocities measured in the depth cell) and assigned to the center of each depth cell over the measured vertical.

Computing Discharge

An ADCP deployed on a moving boat can compute the discharge in real time while traversing the stream from one bank to the other (Simpson and Oltmann, 1993). Unfortunately the ADCP is unable to measure the entire water column (fig. A-4). Near the water surface, an unmeasured zone is associated with immersion of the ADCP into the water, the blanking distance is below the transducer where data cannot be collected, and an additional unmeasured range is dependent on the water-mode configuration and the depth-cell size. The ADCP also cannot measure all the way to the streambed due to the potential for side-lobe interference. As the ADCP approaches a streambank, the depths will eventually get too shallow for valid data collection. Therefore, the discharge computed by an ADCP is a summation of the measured por-tion of the cross section and extrapolated discharge estimates for unmeasured portions of the cross section at the top, bottom, and both banks. The computation of the measured, top, and bottom portions of the cross section occur for each ensemble, and the discharge for the edges are added to the total.

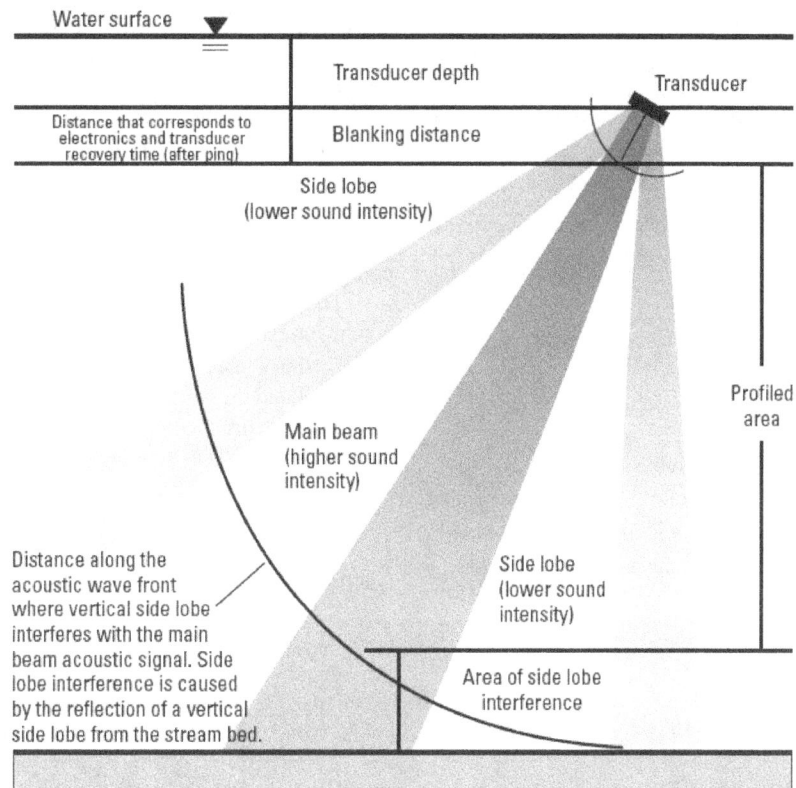

Figure A-4. Acoustic Doppler current profiler beam pattern and locations of unmeasured areas in each profile (from Simpson, 2002).

$$Q = Q_{LeftEdge} + Q_{Top} + Q_{Measured} + Q_{Bottom} + Q_{RightEdge}, \quad \text{(A10)}$$

where

Q	is the total discharge;
$Q_{LeftEdge}$	is the discharge estimated for the unmeasured area near the left bank;
Q_{Top}	is the discharge estimated for the top unmeasured area;
$Q_{Measured}$	is the discharge measured directly by the ADCP;
Q_{Bottom}	is the discharge estimated for the bottom unmeasured area; and
$Q_{RightEdge}$	is the discharge estimated for the unmeasured area near the right bank.

The algorithms commonly used to compute discharge in each portion of the cross section are presented below.

Measured Discharge

Traditional point-velocity meters (Rantz and others, 1982) compute discharge as the product of the cross-sectional area and the mean water velocity perpendicular to the cross-sectional area.

$$Q = A\overline{V}. \quad \text{(A11)}$$

where

A	is the cross-sectional area, and
$\overline{V}$	is the mean water velocity perpendicular to the cross-sectional area.

The ADCP discharge computation algorithm is based on this same principle. Figure A-5 illustrates the water and boat velocity vectors for a single depth cell in an ADCP

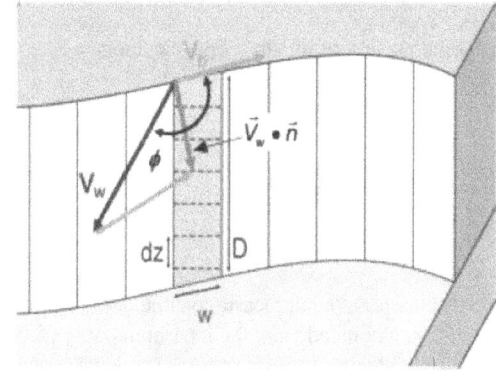

Figure A-5. Boat- and water-velocity vectors for a single bin in an ADCP ensemble.

transect. The cross-sectional area of the depth cell can be computed as the product of the depth-cell size (dz) and ensemble width (W). The depth-cell size is defined by the instrument configuration. The ensemble width is computed as the product of the boat speed and time between ensembles.

$$A = Wdz = \left|\vec{V}_b\right| dt dz \qquad (A12)$$

where

$\left|\vec{V}_b\right|$ is the magnitude of the boat velocity vector, and

dt is the time between ensembles.

The discharge for the first ensemble is zero because to compute the ensemble width, a time for the previous ensemble is needed to compute dt. The water velocity can be computed as the product of the magnitude of the water velocity, $\left|\vec{V}_w\right|$, and the unit normal vector of the boat velocity, $\vec{n}$.

$$\overline{V} = \left|\vec{V}_w\right| \bullet \vec{n} \qquad (A13)$$

The equation for discharge in a depth cell, Q_{bin}, becomes

$$Q_{bin} = \left(\left|\vec{V}_w\right| \bullet \vec{n}\right)\left|\vec{V}_b\right| dt dz. \qquad (A14)$$

Applying trigonometric relations, equation A14 can be written as

$$Q_{bin} = \left|\vec{V}_w\right|\left|\vec{V}_b\right| \sin\phi \, dt dz. \qquad (A15)$$

The product of the vector magnitudes and sine of the internal angle is defined as the vector cross product, and the equation for Q_{bin} can be written in terms of the water- and boat-velocity vector components:

$$Q_{bin} = (\vec{V}_w \times \vec{V}_b) dt dz = (V_{wx}V_{by} - V_{wy}V_{bx}) dt dz. \quad (A16)$$

The measured portion of the discharge can then be computed as

$$Q_{Measured} = \sum_{j=1}^{Ensembles} \sum_{i=1}^{Bins} Q_{bin}. \qquad (A17)$$

Top Discharge

The discharge in the unmeasured zone at the water surface must be estimated from the measured data. The length of the unmeasured zone at the water surface is determined by the draft of the instrument deployment, the blank required to avoid the effects of ringing, the water mode configuration,

and depth-cell size (see the Limitations of ADCPs section for a detailed discussion of ringing). The methods for estimating the top discharge can be applied to the individual velocity components (approach used by SonTek/YSI) or to the cross product from equation A16 (approach used by TRDI). Both approaches are mathematically identical. For simplicity, only the equations using the cross product are shown herein. See Huang (2000) for equations based on individual velocity components.

The simplest assumption for estimating the top discharge is to assume the velocity (cross product) in the topmost valid depth cell is a good estimate of the mean velocity between that depth cell and the water surface. This is typically referred to as the constant extrapolation method:

$$Q_{Top} = \sum_{j=1}^{Ensembles} \chi\left(z_{ws}^{b+1} - z_{tb}^{b+1}\right) dt, \qquad (A18)$$

where

χ is the velocity cross product,

z_{ws} is the range from the streambed to the water surface, and

z_{tb} is the range from the streambed to the top of the topmost valid depth cell.

This constant extrapolation method is often used where there is an upstream wind or an irregular velocity profile through the measured portion of the water column.

A valid but slightly more complicated approach is to assume that the water-velocity profile follows a logarithmic distribution. Simpson and Oltmann (1993) attempted to calculate discharge using several different methods for profile estimation (logarithmic and general power law), but found that because of "noisiness" of the ADCP-profile data, the resulting least-squares-derived estimates were unrealistic, especially near the upper part of the profile. A method using a one-sixth power law (Chen, 1989) eventually was chosen because of its robust noise rejection capability during most streamflow conditions.

$$\chi = a z^b, \qquad (A19)$$

where

a is a coefficient derived from a least-squares fit of the equation to the measured data,

z is the range from the streambed to the location of the value of χ, and

b is the exponent commonly assumed to be 1/6.

The power-law estimation scheme is an approximation only and emulates a Manning-like vertical distribution of horizontal water velocities. Different power coefficients can be used to adjust the shape of the curve fit to emulate profiles measured in an estuarine environment or in areas that have bedforms that produce nonstandard hydrologic conditions. Appling equation A19 and integrating to obtain the top discharge estimate yields

$$Q_{Top} = \sum_{j=1}^{Ensembles} \frac{a}{b+1}\left(z_{ws}^{b+1} - z_{tb}^{b+1}\right)dt. \qquad (A20)$$

Both SonTek/YSI and TRDI provide additional options to estimate the top discharge for nonstandard profiles. SonTek/YSI allows the user to select the number of depth cells used in the top extrapolation method. TRDI restricts the constant method to use of the top depth cell only and the power-law estimate to using all of the available depth cells, but provides an additional 3-point slope method to fit situations where wind significantly affects the velocity at the water surface. The user is referred to the manufacturer's documentation for more detail on these methods (SonTek/YSI, 2007; Teledyne RD Instruments, 2007).

Bottom Discharge

ADCPs cannot measure the water velocity near the streambed due to side-lobe interference (see the Limitations of ADCPs section for a detailed discussion of side-lobe interference). Unlike the top discharge estimation problem where the velocity at the water surface is not known, we have some understanding of the water velocity at the streambed. From fluid mechanics, it is known that the water velocity must go to zero at the streambed and that a logarithmic velocity profile is a reasonable approximation of the velocity profile in the boundary layer (Schlichting, 1979). Therefore, the power law is always used to compute the discharge in the bottom unmeasured portion of the water column.

$$Q_{Bottom} = \sum_{j=1}^{Ensembles} \frac{a}{b+1}z_{bb}^{b+1}dt, \qquad (A21)$$

where

z_{bb} is the range from the streambed to the bottom of the bottom most valid depth cell.

To better apply this method to situations where profile throughout the water column may not follow a logarithmic distribution, such as for bidirectional flow, SonTek/YSI allows the user to select the number of depth cells near the bottom of the water column that are used in the least-squares determination of a. TRDI adds an additional method to their software called "No-Slip," which applies equation A21 but restricts the least-squares determination of a to depth cells in the bottom 20 percent of the profile, or in the absence of valid depth cells in the bottom 20 percent, the last valid depth cell is used to compute a.

Edge Discharge

The unmeasured discharge at the edges of the stream are estimated using a ratio interpolation method presented by Fulford and Sauer (1986), which can be used to estimate a velocity at an unmeasured location between the riverbank and the first or last measured velocity in a cross section (fig. A-6).

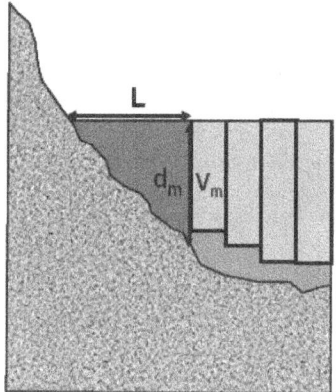

Figure A-6. Components used to calculate estimates of discharge in the unmeasured edges of cross sections.

The equation for the estimate is

$$\frac{V_x}{\sqrt{d_x}} = \frac{V_m}{\sqrt{d_m}}, \qquad (A22)$$

where

x is a location midway between the riverbank and first or last ADCP-measured ensemble or $L/2$,

V_x is the estimated mean velocity at location x,

V_m is the measured mean velocity at the first or last ADCP-measured subsection (fig. A-6),

d_x is the depth at location x, and

d_m is the depth at the first or last ADCP-measured ensemble (fig. A-6).

Fulford and Sauer (1986) defined d_m and V_m as depth and velocity, respectively, at the center of the first or last measured subsection and not the near-shore edge of the ensemble, as presented in equation A22. However, because the ADCP ensembles purposely are kept very narrow at the start and finish of each measurement, the differences between the two applications are not significant (Simpson and Oltmann, 1993).

With the Fulford and Sauer (1986) method, discharge can be estimated by assuming a triangular area at the edge as

$$Q_{Edge} = A_{edge}V_{L/2} = 0.5Ld_m * V_m \frac{\sqrt{0.5d_m}}{\sqrt{d_m}} = 0.3535Ld_mV_m, \quad \text{(A23)}$$

where

Q_{Edge} is the estimated discharge in the unmeasured edge,

A_{edge} is the area of the unmeasured edge,

$V_{L/2}$ is the velocity midway between the bank and the first or last ADCP-measured ensemble, and

L is the distance from the last valid ensemble to the edge of water (fig. A-6).

Equation A23 can be written in a more general form, which utilizes an edge-shape coefficient.

$$Q_{Edge} = C_eV_mLd_m, \quad \text{(A24)}$$

where

C_e is an edge-shape coefficient.

The edge-shape coefficient can be adjusted by the user to reflect unusual edge shapes or roughness but is commonly set to 0.3535 for triangular edges and 0.91 for rectangular edges. The value for L must be measured and entered by the user (see discussion of laser range finders in the Other Equipment section). It is also recommended that 5 to 10 seconds of data be collected in a near-stationary position at the beginning and end of each transect to obtain a good measurement for V_m and L.

Appendix B – Collecting Data in Moving-Bed Conditions

Cause and Effect of a Moving Bed

To measure absolute water velocities with an ADCP (water velocities relative to a fixed reference), the ADCP must sense and measure the velocity of the ADCP itself, relative to the river bottom. This is referred to as bottom tracking. Bottom-tracking measurements are similar to water-velocity measurements, but separate acoustic pulses are used. If the velocity of the water is known relative to the ADCP (water tracking), and the velocity of the ADCP is known relative to the river bottom (bottom tracking), then the water velocity relative to the bottom can be calculated. However, for the calculation of actual water velocities to be accurate, the streambed must be stationary.

The bottom-track pulse often is longer than the water-track pulse to allow complete ensonification (the filling of any fluid medium with acoustic radiation, which is then observed and analyzed to study the medium or to locate or image objects within it) of the bottom. For the bottom ping to accurately measure the depth to and Doppler shift from the streambed, the pulse should uniformly ensonify the surface of the streambed so that it will receive uniform backscatter. If the pulse is too short, the echo returns first from the part of the beam closest to the ADCP, followed by successively further areas (fig. B-1). The beam angle is different for the near and far parts of the beam, making analysis of the signal difficult. A pulse ensonifying the entire bottom will produce an accurate and stable measurement of the velocity of the instrument. However, sediment transport on or near the streambed can affect the Doppler shift of the bottom-tracking pulses and can cause a measurement error referred to as a moving bed. When the bottom is ensonified the sediment above the bottom also is ensonified and included in the backscattered signal (fig. B-2); therefore, the backscattered signal of the long pulse may be biased by the sediment in the water column just above the streambed. In such situations, reflections of bottom-tracking pulses from highly concentrated near-bed sediments are difficult to distinguish from reflections from the bed. These near-bed sediments typically are being transported in the downstream direction. If bottom tracking is affected by sediment transport, the measured boat velocity will be biased in the opposite direction of the sediment movement, which would make a stationary boat appear to be moving upstream (fig. B-3). Therefore, if sediment moving near the streambed is measured by bottom-tracking measurements, the boat will have an apparent upstream velocity, the calculated downstream water velocity will be reduced, and the corresponding discharge will be biased low.

Determining whether a moving-bed condition exists at a site is not always intuitive to a hydrographer because a moving bed may be measured by an ADCP even in streams where the bed is visually stationary. The detection of a moving bed also is dependent upon the frequency of the ADCP. Higher frequency ADCPs are more sensitive to sediment transport

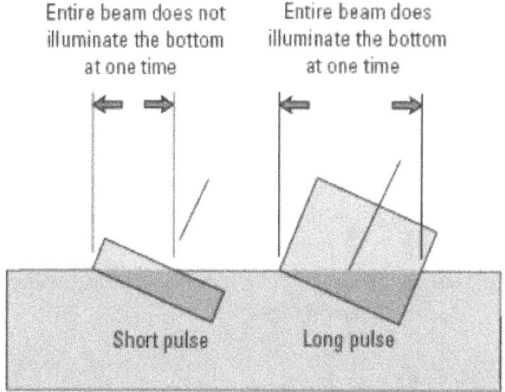

Figure B-1. An example of short and long bottom track pulses (from Simpson, 2002).

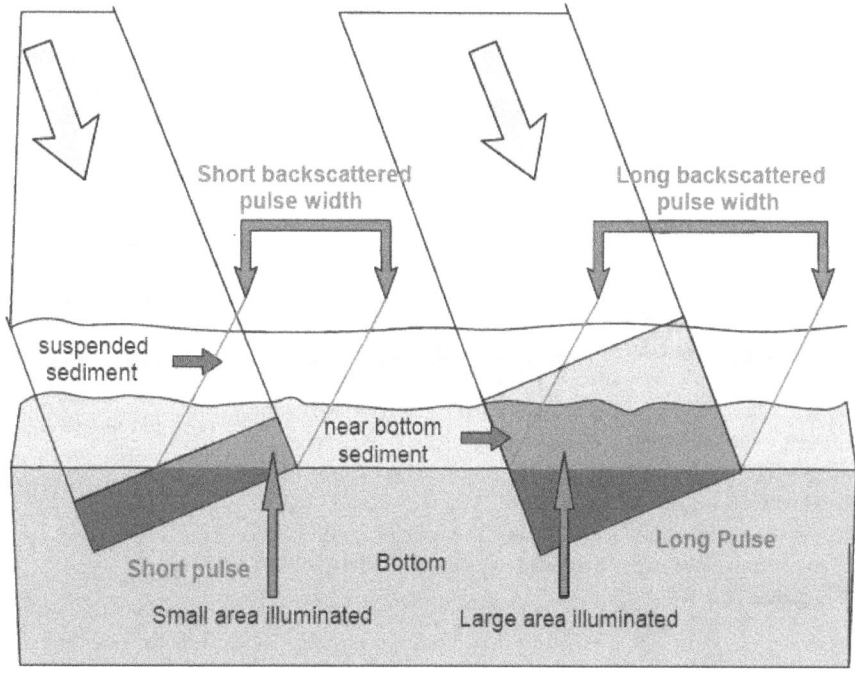

Figure B-2. An example of the potential for water bias (moving bed) for short and long bottom track pulses (from Simpson, 2002).

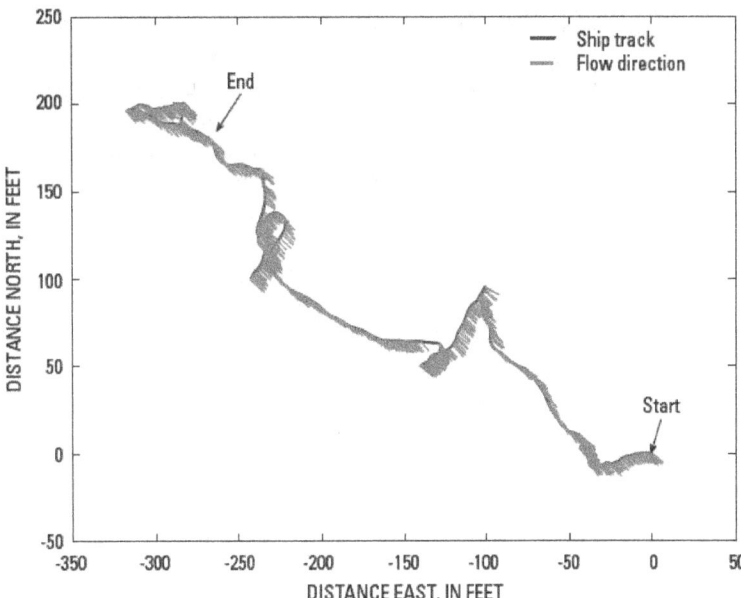

Figure B-3.   A moving bed measured with a 1,200-kilohertz acoustic Doppler current profiler (ADCP) on the Mississippi River at Chester, Illinois.

than are lower frequency ADCPs and, therefore, are more prone to detecting a moving bed.

Methods to Identify a Moving Bed

USGS policy (U.S. Geological Survey, 2002b; Oberg and others, 2005) requires that a moving-bed test be conducted prior to making a discharge measurement. In USGS training classes and in Oberg and others (2005), an exception to conducting a moving-bed test with every measurement was allowed if sufficient documentation was provided in the field folder to justify not conducting a moving-bed test. Field experience has shown that sediment-transport characteristics can vary greatly for the same discharge, depending on the hydrograph shape, source of runoff, and season of the year. Thus, a moving bed may be detected at a location and discharge where one was not previously detected, or a moving bed may no longer exist at a location and discharge where one previously was detected. Moving-bed conditions also have been observed in low-velocity environments (less than 1 ft/s) and likely are caused by organic material transported by the water. To ensure the quality of the data collected, every moving-boat measurement made with an ADCP must have a recorded moving-bed test. If a site routinely has a moving bed and GPS is always used with the ADCP, a moving-bed test is still required but need only be 5 minutes in length. This requirement supersedes Oberg and others (2005) and various USGS training materials.

Stationary Test with No GPS

The first moving-bed-test method requires that the boat with the ADCP be held in a stationary position while recording ADCP data, using bottom tracking as the boat-velocity reference. If the stationary position is maintained by a tether or anchor so that upstream or downstream movement of the ADCP is not possible, the moving-bed test should be recorded for no less than 5 minutes; however, if the ADCP can move either upstream or downstream, such as when the boat operator is trying to maintain position of the boat, the test should be recorded for no less than 10 minutes. These criteria supersede the guidance on station-ary moving-bed tests that have been previously published in Oberg and others (2005) and U.S. Geological Survey (2002b). When a moving-bed condition is present, a stationary boat will appear to have moved upstream (fig. B-3). The error caused by the moving bed can be estimated by dividing the distance of the apparent boat motion in the upstream direction by the duration of the test in seconds. This computation will provide the moving-bed velocity detected by the bottom-tracking technique. This moving-bed velocity can then be divided by the average water velocity from the moving-bed test and multiplied by 100 to yield an estimate of the percent bias error for a water-velocity measurement at this stream location. In the example shown in figure B-3, the distance traveled was approximately 360 ft for a 10-minute (600-second) period. The estimated moving-bed velocity is 0.59 ft/s. If the mean velocity for the discharge measurement was 4.9 ft/s, the percent bias error in the water-velocity measurement would be estimated to be 12 percent. Criteria for determining if a moving bed is present must account for the accuracy of the test. If the moving-bed test was completed with a fixed tethered deployment, an anchored manned boat, or a manned boat where the user is sure there was little movement of the boat, a moving bed is determined to be present when the measured moving-bed velocity is greater than 1 percent of the mean water velocity at the test location. If the moving-bed test was made using a manned boat that was not anchored and may have moved either upstream of downstream a criteria of 2 percent instead of 1 percent is used because uncertainty has been introduced into the test by boat movement. Discharge-measurement techniques that are not affected by a moving bed, or that correct for the effect of a moving bed, should be used if a moving bed has been detected.

When it is not possible to safely anchor the manned boat in the stream because of boat traffic, drift, or other hazardous conditions, holding the boat stationary at the desired location in the measurement section may be difficult. Boat movement in the upstream or downstream direction will introduce errors in the moving-bed test when using this method. A technique for helping the boat operator determine whether the boat is moving excessively in the upstream or downstream direction during a moving-bed test is illustrated in figure B-4. The boat operator selects two distinguishable reference points on shore that are separated by a considerable distance from one another (98 ft or more). Examples of distinguishable reference points include telephone or power poles or large trees that can be easily distinguished from other nearby trees. Bridges, bridge piers, and navigation buoys also can be used as reference points for maintaining an approximately stationary position when making a moving-bed test. If the boat changes position appreciably in the upstream or downstream direction, this change should be noted on the field note form, and the test should be repeated, if possible.

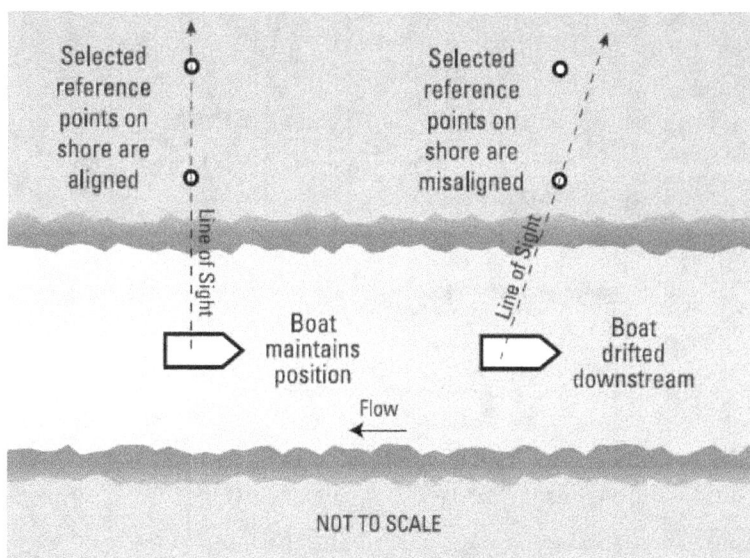

Figure B-4.   Method for verifying that upstream/downstream boat movement is minimized during an acoustic Doppler current profiler moving-bed test (modified from Environment Canada, 2004).

The TRDI StreamPro ADCP does not include an internal compass. The cross product of the water-velocity and boat-velocity vectors, used for the computation of discharge, does not depend on the heading of the instrument (Simpson and Oltmann, 1993). However, users began reporting moving-bed biases when using the StreamPro where it was apparent from the velocity and bed material that little or no sediment transport was occurring. The false moving bed observed in stationary moving-bed tests conducted with a StreamPro is the result of rotation and lateral movement of the StreamPro on the end of the tether. During a stationary moving-bed test the StreamPro and float often tend to swim or kite from the end of the tether. Because the StreamPro does not have an internal compass, it cannot compensate for instrument rotation. The lack of a compass in a StreamPro does not allow the StreamPro to report boat or water velocities referenced to a fixed rotational reference, such as magnetic north; rather, the StreamPro reports velocities in instrument coordinates.

Careful analysis of false moving-bed ship tracks indicate that ship movement (kiting or swimming) typically is perpendicular to the water velocity. When the StreamPro rotates, the boat and water velocities are rotated. Thus, the boat velocity and water velocities remain in the correct orientation relative to each other. This characteristic permits the discharge to be computed without a compass. This characteristic also can be used to determine the velocity of the boat in the upstream direction. The boat velocity in the upstream direction measured during a stationary moving-bed test can be computed using the vector dot product of the boat velocity and water velocity. The use of the dot product for analysis of a moving-bed test was originally suggested by Randy Marsden (Teledyne RD Instruments, oral commun., 2006). This approach subsequently has been developed and applied to the false moving-bed problem associated with the StreamPro ADCP. The USGS has developed software (Stationary Moving-Bed Analysis (SMBA)) that applies the dot-product approach and provides a simple and efficient user interface to allow users to apply the dot-product approach to all StreamPro stationary measurements, which has been shown to eliminate the effect of the boat/instrument motion on the stationary moving-bed test. The SMBA program can be obtained from the USGS at *http://hydroacoustics.usgs.gov/movingboat/ mbd_software.shtml*.

Stationary Test with GPS

A more accurate method for estimating the errors introduced by a moving bed can be determined if a GPS is available for use and is interfaced with the ADCP and the data-collection software. The stationary test with GPS method also requires that the ADCP boat be held in a stationary position and a data file recorded for at least 5 minutes, if GPS data are of good quality. Either bottom tracking or GPS can be used for the boat-velocity reference. Using GPS as the reference is often easier during data collection so that the boat operator can use the ship-track display as a means to assist in holding the boat in an approximately stationary position. After the moving-bed test is complete, compare the boat track using ADCP bottom track as reference with the boat track using GPS as the reference. If the bottom-track boat velocity data indicate apparent upstream movement that the GPS data do not indicate, a moving bed is present. The error caused by the moving bed can be computed in the same manner as described for the stationary test with no GPS method, except that the distance in the upstream direction indicated by bottom tracking should be corrected by the distance actually traveled in that direction, as indicated by GPS (Oberg and others, 2005). In the WinRiver software, this distance can be found in the "compass calibration" tabular window and is labeled "BMG-GMG mag," and the direction of the "BMG-GMG dir" should be in the upstream direction. If the measured moving-bed velocity is greater than 1 percent of the mean water velocity at the test location, discharge-measurement techniques that are not affected by a moving bed, or that correct for the effect of a moving bed, should be used.

Loop Method

If the ADCP can be held stationary, stationary moving-bed tests are a good measure of the magnitude of an apparent moving streambed; however, these tests represent moving-bed conditions for only one location in the cross section. An alternative to the stationary moving-bed test is the loop method, which is based on the fact that as an ADCP is moved across the stream, a moving bed will cause the bottom-track-based ship track to be distorted in the upstream direction. Therefore, if an ADCP makes a two-way crossing of a stream (loop) with a moving bed and returns to the exact starting position, the bottom-track-based ship track will show that the ADCP appears to have returned to a position upstream from the original starting position (fig. B-5). The mean moving-bed velocity can be estimated from the distance the ADCP appeared to have moved upstream from the starting position (loop-closure error) and the time required to complete the loop.

$$\bar{V}_{mb} = \frac{D_{up}}{T}, \qquad (B1)$$

where

$\bar{V}_{mb}$ is the mean moving-bed velocity for the measurement section,

D_{up} is the loop-closure error (distance made good, straight-line distance from starting point to ending point), and

T is the measurement time required to complete the loop.

These data are readily available from most commercial software used to measure discharge with ADCPs (fig. B-6).

The loop method must be applied properly, or it may produce incorrect results. Anyone planning to use the loop method should read and follow USGS Scientific Investigations Report 2006–5079 (Mueller and Wagner, 2006), which describes the procedures, limitations, and uncertainties associated with the loop method. The following steps are for collecting and analyzing loop-test data.

1. *Calibrate the ADCP/ADP (acoustic Doppler profiler) compass using internal calibration routines.* A compass calibration accuracy of better than 1 degree is desired. Calibrations with errors greater than 1 degree should be repeated. If a calibration of less than 1 degree cannot be

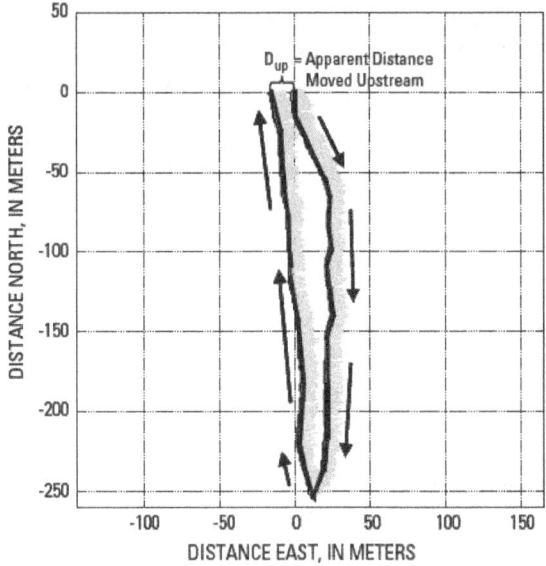

Figure B-5. A distorted ship track in a loop caused by a moving bed.

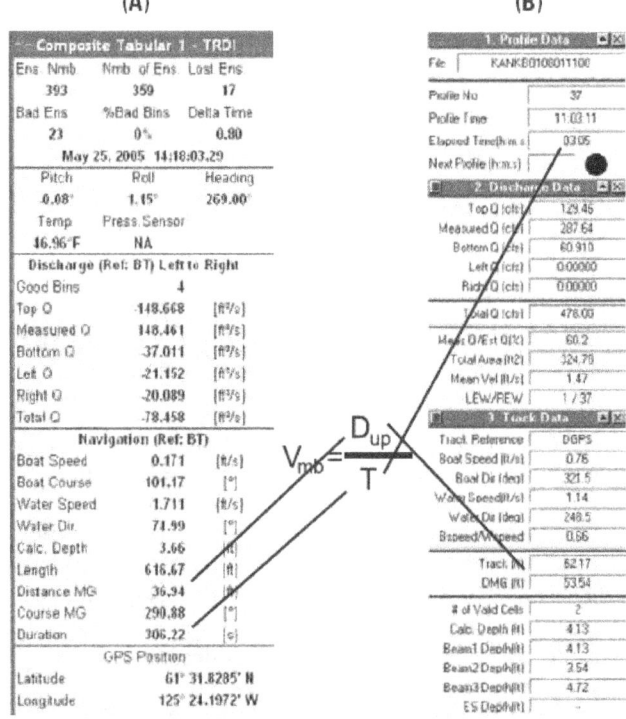

Figure B-6. Example of parameters used to compute the mean correction for data collected with an ADCP and displayed with (A) WinRiver and (B) RiverSurveyor.

obtained after several attempts, appropriate field notes should be recorded to document the problem. Compass errors greater than 1 degree result in increased errors in the loop-method correction.

2. *Establish a marked starting point where the ADCP/ADP can be returned to the exact location.* This point is not required to be as near to a bank as the end of a regular transect. For example, with a tethered boat, controlling the boat at the edge of the bank can be difficult because of conditions such as slack water, eddies, or vegetation; therefore, establishing a point farther out in the flow could make navigating the boat back to the starting point more practical. Use of a buoy or other fixed object is recommended.

3. *Make a steady pass back and forth across the stream as a normal discharge measurement, but do not stop recording at the far bank.* At the starting point, make sure the boat is ready to begin the transect before beginning to record. **A uniform boat speed is important.** Do not spend extra time at the edges. Plan the loop so that a smooth change in boat direction can be achieved near the far bank. Too much time near the banks will result in a low bias.

4. *Maintain the proper boat speed.* The recommended maximum boat speed should be the lesser of a boat speed that requires no less than 3 minutes to complete the loop or a boat speed that is less than 1.5 times the mean water speed.

5. *Return to the starting point.* Accurately returning to the starting point is very important.

6. *Process the loop file to the end.* Record the distance made good (DMG) and the time required to complete the loop. Note: The DMG in a moving-bed condition should be in the upstream direction (fig. B-5). If the primary direction of the DMG is in a direction other than upstream, this distance may be the result of compass or bottom-track errors, and no moving bed will be assumed.

7. *Compute the mean moving-bed velocity.*

$$\overline{V}_{mb} = \frac{D_{up}}{T}, \tag{B2}$$

where

$\overline{V}_{mb}$ is the mean velocity of the moving bed,
D_{up} is the DMG, and
T is the measurement time required to complete the loop.

8. *Compute the ratio of the mean moving-bed velocity to the mean water velocity.* Note: Because the loop closes on itself, the net discharge and cross-sectional area approach zero; thus, the mean velocity should be the mean of the measured velocities and not the discharge divided by the cross-sectional area.

9. *Determine if the ratio exceeds the recommended criteria.* In order to minimize the potential error in the measured

moving-bed bias when using the loop method as a moving-bed test, the hydrographer should use the following thresholds in determining and applying a correction for an apparent moving streambed.

- When using the loop method, a measured mean moving-bed velocity of at least 0.04 ft/s indicates the presence of a moving bed.

- If the measured moving-bed velocity exceeds 0.04 ft/s, the ratio of the mean moving-bed velocity and mean water velocity should be computed.

- If this ratio is greater than 0.01, the apparent bed movement will cause at least a 1-percent negative bias in the computed discharge, and a DGPS or method that accounts for or corrects for a moving bed should be used.

Methods to Account for Moving-Bed Effects

The integration of a GPS to measure the velocity of the ADCP has been shown to alleviate the systematic errors associated with a moving bed (Mueller, 2002) and is the most accurate way to measure discharge in moving-bed conditions. If, however, a GPS is not available or site conditions do not permit its use, methods have been established to correct discharge measurements biased by a moving bed. The use of a GPS and the application of various discharge correction methods are discussed in the following sections.

Using GPS with ADCPs

Using GPS with ADCPs eliminates the effect of a moving bed on the velocity measurements but introduces several sources of potential error. Understanding how GPS operates and how it is used with ADCPs is important for collecting high-quality ADCP measurements when using GPS data as the boat-velocity reference.

The computation of water velocity from an ADCP mounted onto a moving boat is a vector-algebra problem. The ADCP measures the water velocity relative to the moving boat (relative water velocity), so the velocity of the boat must be accounted for to obtain the true water velocity. The true water velocity is computed by subtracting the bottom-tracking velocity from the water-tracking velocity. When bottom tracking is used, the direction of the boat velocity vector as measured by bottom tracking (θ_{BT}) and water-velocity vector (θ_{WT}) are referenced to the ADCP (fig. B-7A). The ADCP has an internal fluxgate compass to measure the orientation of the instrument (θ_{Inst}) relative to the local ambient magnetic field (magnetic north). The water-velocity vector can be easily referenced to magnetic north by rotating the vector based on the measured θ_{Inst} and to true north by again rotating the vector by a user-specified magnetic variation (θ_{Mag}). The magnitude of the water velocity is unaffected by any errors in the

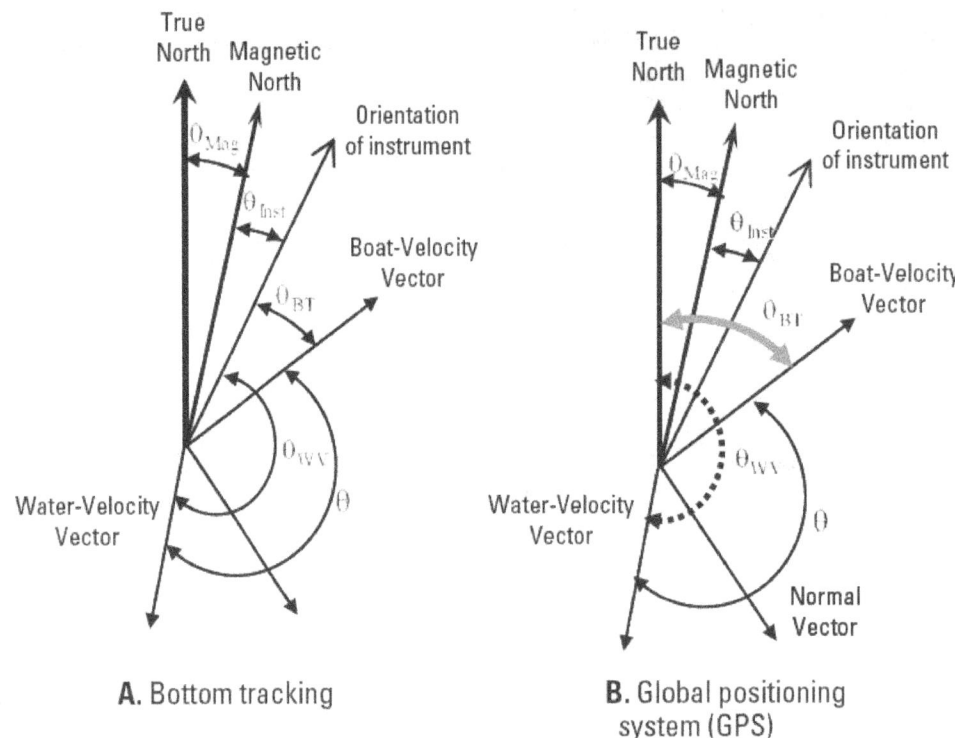

Figure B-7. Boat-velocity vectors referenced by (A) bottom tracking and (B) global positioning system (adapted from Mueller, 2002).

measurement of θ_{Inst} or entry of θ_{Mag} when bottom tracking is used as the boat-velocity reference. The basic equation presented in Simpson and Oltmann (1993) for computing measured discharge (exclusive of unmeasured areas) by use of an ADCP mounted onto a moving boat is

$$Q = \int_0^T \int_0^D |\vec{V}_f| |\vec{V}_b| \sin\theta \, dz \, dt, \qquad (B3)$$

where

Q	is the total discharge,
T	is the total time for which data were collected,
D	is the total depth,
$\vec{V}_f$	is the mean water-velocity vector,
$\vec{V}_b$	is the mean boat-velocity vector,
θ	is the angle between the water-velocity vector and the boat-velocity vector (fig. B-7),
dz	is the vertical differential depth, and
dt	is differential time.

To compute the discharge, only the angle between the water-velocity and the boat-velocity vectors is needed. When GPS is used to determine the boat-velocity vector, this vector is

referenced to true north as determined from the GPS data (fig. B-7B). The orientation of the instrument relative to true north must be determined to put the boat-velocity vector and the relative water-velocity vector in the same coordinate system and allow for the computation of the water-velocity vector and θ. The discharge is affected by errors in measuring θ_{Inst} and in the determination of the magnetic variation (θ_{Mag}) when GPS is used as the boat reference. The errors associated with θ_{Inst} can cause errors in the measured discharge that are proportional to the speed of the boat. Proper setup and calibration of the ADCP's internal compass, determination of the local magnetic variation, and a slow boat speed are critical to quality discharge measurements made using GPS data as the boat-velocity reference.

Errors associated with fluxgate-compass measurements can result from horizontal accelerations of the ADCP or environmental conditions near the instrument. Most fluxgate compasses are gimbal-mounted, which allows them to measure the Earth's horizontal magnetic field. When the instrument is subject to horizontal accelerations, such as when a boat accelerates or turns, the force generated by the acceleration causes the compass to swing out of the vertical position and measure something other than the horizontal magnetic field. Most of the significant errors associated with horizontal accelerations can be eliminated by slow, smooth boat operation.

Errors associated with fluxgate-compass measurements caused by environmental conditions can be classified as one- and two-cycle errors. One-cycle errors are caused by permanent magnets and current-carrying conductors; two-cycle errors are caused by iron and magnetically permeable material. ADCPs manufactured by TRDI and SonTek/YSI for making discharge measurements from a moving boat have firmware routines to allow the calibration of the compasses in place to compensate for environmental conditions.

The local magnetic variation (or declination) can be either estimated or measured, depending upon site conditions. Estimates of the local magnetic variation can be obtained from USGS 7.5-minute quadrangle maps, magnetic field charts, and geomagnetic field models. Although these estimated values are often accurate, some areas have appreciable magnetic anomalies that are not accurately predicted by models or general charts. Teledyne RD Instruments (2003) documents a procedure for measuring the magnetic variation onsite by use of an ADCP and a DGPS. This same procedure can be used with RiverSurveyor instruments and RiverSurveyor software from SonTek/YSI. The limitation of this procedure is that there can be no moving-bed conditions because both the bottom tracking and GPS are used in the computations.

GPS provides two options for determining boat velocity: (1) differentiated position using the GGA NMEA-0183 sentence and (2) Doppler-based velocity reported in the VTG NMEA-0183 sentence. The GGA data sentence broadcast by the GPS includes time, positional data (latitude, longitude, and elevation), and information about the satellite constellation used to reach the position solution. When using the GGA sentence from the GPS to measure the movement of the ADCP, the velocity of the instrument is determined by computing the distance traveled between successive GPS position solutions and dividing that distance by the time between the solutions. Hence, positional accuracy is vitally important to achieve an accurate measure of ADCP movement using the GGA sentence; therefore, a differential correction signal is required. To use the GGA sentence, DGPS receivers are required, and receivers should have a 95-percent accuracy of about 3.3 ft or less in the horizontal location.

Although the differential correction accounts for errors induced by the ionosphere, atmosphere, and selective availability, the user must be aware of and take action to minimize uncorrectable errors, which can be caused by the user, the satellite configuration, or the characteristics of the site. Place the GPS antenna as near to the center of the ADCP as possible so that the direction of travel is the same for both the antenna and the ADCP during all boat maneuvers. The antenna should be located above the boat cabin or other accessories on the boat to eliminate multipath errors. Multipath errors are positional errors that arise when satellite signals bounce around before getting to the receiver, rather than taking a direct path. The result is a barrage of signals arriving at the receiver: first the direct one, then a series of delayed reflected ones. This creates a noisy signal, and if the bounced signals are strong enough, they can confuse the receiver and cause erroneous positional measurements. Sophisticated receivers use a variety of signal processing techniques to make sure they only consider the earliest arriving signals, which are the direct ones. Although these techniques reduce multipath errors, they do not completely eliminate the problem.

Occasionally, the configuration of the satellites does not allow an accurate determination of the horizontal position. This can be monitored using the horizontal dilution of precision (HDOP). If the HDOP parameter is greater than 2, or if the HDOP changes by more than 1 during a transect, the quality of the DGPS positions is suspect. Local site characteristics, such as canyon walls, bluffs, tall buildings, and tree cover, can result in poor DGPS positions because of multipath errors and loss of satellite visibility. Poor satellite visibility often results in numerous changes in the number and configuration of satellites used to determine a position. Numerous changes in satellites are another indication that the quality of the DGPS positions may be poor. In addition to horizontal-position coordinates, the DGPS also computes elevation. This elevation is 2 to 4 times less accurate than the horizontal position (Mueller, 2002). The elevation of the boat should be reasonably constant. Changes greater than 11.5 ft in the DGPS-determined elevation indicate that the quality of the DGPS positions may be poor (Teledyne RD Instruments, WinRiver 10.06 Help File, written commun., 2006).

While GPS is most often used to determine positions, many GPS receivers also can be used to measure velocity relative to ground with an assessment of the Doppler shift in the satellite carrier phase frequencies, which is typically reported in the VTG sentence. The method uses the actual signal frequency, and not a phase angle, to determine the Doppler shift. No ambiguity can be present in the computed velocities. As for the position determination, the velocity measurement requires the use of at least four satellites. The quality of the solution also is influenced by the number of satellites and the shape of the constellation (HDOP) during the observation. This alternative has an advantage because it is minimally affected by multipath and satellite changes because of the short sampling time required. In addition, multipath and ionospheric/atmospheric distortions do not affect the precision of the measurement. As a result, the Doppler measurement of velocity can be produced without the need for any differential correction. Emerging research by Environment Canada and the USGS using the VTG sentence from a GPS has shown the potential to provide better measurements of ADCP movement than the GGA data string and without the need for differential correction (Francois Rainville, Water Survey of Canada, written commun., 2007). Additional research is underway to investigate the limitations of the Doppler-derived velocity in the VTG sentence.

Alternatives to Using a GPS

Using a DGPS to measure the velocity of an ADCP has been shown to alleviate the systematic errors associated with a moving bed (Mueller, 2002). GPS systems, however, do not work in all conditions. For example, a GPS will have trouble providing consistently accurate positions and velocities on waterways with dense tree canopy along the banks, in deep valleys or canyons, and near bridges because of multipath and satellite signal reception problems. Alternative methods of correcting for the systematic moving-bed error have been developed for situations where integrating GPS with an ADCP is not possible.

Loop Method

The results of a loop method moving-bed test (previously described) can be used to compute the discharge missing from the measurement caused by the moving bed. This computed missing discharge can be added to the measured discharge to yield a corrected discharge.

$$Q_{TC} = Q_{TM} + Q_{mb}, \qquad \text{(B4)}$$

where

$\quad Q_{TC} \quad$ is the discharge corrected for the moving-bed bias,

$\quad Q_{TM} \quad$ is the measured discharge, and

$\quad Q_{mb} \quad$ is the discharge not measured because of the moving bed.

A detailed analysis of the application of the loop method was conducted by the USGS (Mueller and Wagner, 2006).

Careful field procedures are absolutely critical to the successful application of the loop method. Failure to accurately return the instrument to the starting point, an uncalibrated or improperly calibrated compass, or loss of bottom track during the loop will result in unpredictable errors that render this technique unusable. Current research, which is limited by the amount of available field data, indicates that site-specific characteristics and data-collection techniques, such as the shape of the measurement section, distribution of the moving-bed velocity, time spent at the banks, boat speed, and uniformity of the boat speed, can affect the discharge correction by 10 percent or greater. When applied properly, however, this technique should consistently yield total corrected discharges that are within 5 percent of the actual discharge. *The loop method cannot be applied with a TRDI StreamPro ADCP because it does not have a compass.*

After completion of a loop method moving-bed test, two methods described by Mueller and Wagner (2006) can be used to correct discharge biased by a moving bed—the mean correction method and distributed correction method. The mean correction method is simple and can be completed with manual computations. The distributed correction method accounts for the velocity distribution and shape of the cross section and requires a computer program.

Mean Correction Method

A simple method for computing the discharge not measured because of the moving bed is to compute the mean moving-bed velocity and multiply it by the cross-sectional area (measured perpendicular to the flow).

$$Q_{mb} = \overline{V}_{mb} A, \qquad \text{(B5)}$$

where

$\quad \overline{V}_{mb} \quad$ is the mean velocity of the moving bed, and

$\quad A \quad$ is the cross-sectional area perpendicular to the mean flow direction.

The mean moving-bed velocity can be computed from the distance the ADCP appeared to have moved upstream from the starting position (loop-closure error) and the time required to complete the loop. These data are readily available from most commercial software used to measure discharge with ADCPs. The cross-sectional area must be computed ***perpendicular to the mean flow direction***. If the cross-sectional area is computed parallel to the ship track measured by the ADCP, then the cross-sectional area will be computed based on a ship track that is distorted in the upstream direction by the moving bed. The distortion of the ship track by a moving bed will result in a cross-sectional area that is too large. The mean correction method is simple to apply but does not account for the cross-section shape and spatial correlation of the sediment transport with the spatial distribution of the discharge in the cross section. Therefore, streams with high spatial variability in sediment transport and discharge distributions may not be properly represented by using a single mean moving-bed velocity to correct the measured discharge.

Distributed Correction Method

The actual moving-bed velocity at any point in the stream is unknown, but the moving-bed velocity is assumed to be proportional to the near-bed water velocity (Callede and others, 2000). The distributed correction method uses a $1/6^{th}$ power curve to provide a consistent estimate of the near-bed velocity at any point in the cross section. To determine the distributed loop method correction, the measured mean moving-bed velocity from the loop is distributed to each ADCP profile by a ratio of near-bed velocity for each profile and the mean near-bed velocity for the cross section. The distributed moving-bed velocities are then applied to the water and boat velocities for all depth cells in each of the corresponding profiles in the measured portion of the cross section to determine the corrected measured discharge. The total discharge measured by an ADCP consists of a measured portion and estimates of discharge in the unmeasured top, bottom, left, and right edges. The final corrected measured discharge is computed using the ratio of the corrected and uncorrected measured portion of the discharge to correct the sum of the measured and top and bottom estimated discharges. Water velocities near the bank are assumed to be sufficiently low as to not cause a moving bed; therefore, no correction is applied to the left and right edge discharges.

Distribution of the mean moving-bed velocity based on near-bed velocities requires a consistent method of determining near-bed velocities at each measured vertical. Due to side-lobe interference the lower 6–10 percent, approximately, of each velocity profile is unmeasured. In addition, invalid velocity measurements are common in the lower portions of the profile. Simply using the last valid velocity in each measured velocity profile would result in near-bed velocities at various distances from the streambed. The 1/6th power law has been shown to be consistent with a logarithmic velocity profile and is commonly used to estimate the unmeasured top and bottom discharges for ADCP measurements (Chen, 1989; Simpson and Oltmann, 1993). The near-bed velocity is computed by fitting the 1/6th power law through zero at the bed and through the mean velocity of the last two valid velocity measurements in the profile.

The computations associated with the distributed correction are best performed using a computer program. A program, LC, has been developed to perform these computations. LC reads ASCII files that are readily output by standard vendor-supplied ADCP software, thus allowing all the utilities of the data-collection and processing software to be used to validate the measured discharge before applying any corrections. The LC program prompts the user for the ASCII output filename that contains the loop data and computes the magnitude and direction of the DMG from the starting and ending points of the loop. If the direction of the DMG is +/– 45 degrees from the upstream direction and the magnitude is greater than the previously stated thresholds for a moving-bed correction, then a correction is recommended. The program then reads and processes all transects specified by the user and applies the method described herein to each transect. The program then computes a corrected discharge for each transect and the corrected mean discharge for the whole measurement. The LC program can be obtained from the USGS at *http:// hydroacoustics.usgs.gov/movingboat/mbd_software.shtml.*

Multiple Moving-Bed Test Method

The second GPS alternative requires the hydrographer to conduct multiple stationary moving-bed tests across the stream. The advantages to using the multiple moving-bed tests are (1) GPS is not required, (2) no compass calibration is required, and (3) the method can be used where bottom track cannot be adequately maintained to apply the loop method. The disadvantages of this method are (1) multiple moving-bed tests are required, (2) ADCP movement is not corrected during the moving-bed tests, (3) the procedure is time consuming, (4) an accurate cross-sectional area projected perpendicular to the mean flow direction is required for all methods except when using the Stationary Moving-Bed Analysis (SMBA) software, and (5) although the discharges are corrected, measured velocities are still uncorrected (biased low).

Field Procedures

1. Conduct at least three stationary moving-bed tests, and note the location of each test by measuring the distance from shore. Additional moving-bed tests will result in

a more accurate discharge correction. Be sure to record these tests and the locations of each stationary test in the cross section.

2. Each moving-bed test should be 5 to 10 minutes in duration. For details, see the Stationary Test with No GPS section under the Methods to Identify a Moving Bed section.

3. Make a typical ADCP discharge measurement.

Processing Procedures

The corrected discharge can be computed from the multiple moving-bed tests by using one of three computation methods: (1) subsection method, (2) average moving-bed velocity method, and (3) distributed method using the SMBA software.

Subsection Method

1. Playback each moving-bed test and determine the mean moving-bed velocity ($\overline{V}_{mb}$) for each section, which can be computed from the distance the ADCP appeared to have moved upstream from the starting position and the time required to make each stationary moving-bed test.

$$\overline{V}_{mb} = \frac{D_{up}}{T}, \tag{B6}$$

where

D_{up} is the DMG in the upstream direction (straight-line distance from starting point to ending point), and

T is the time required to complete the test.

2. Process the first transect and manually subsection it such that the moving-bed tests are located near the middle of each subsection.

3. Determine the cross-sectional area perpendicular to the mean flow direction (A_i) for each subsection. The cross-sectional area will be affected by the moving bed unless corrected to be perpendicular to the mean flow direction.

4. Compute the corrected discharge for the transect

$$Q_{corrected} = Q_{measured} + \sum (\overline{V}_{mb_i} A_i), \tag{B7}$$

where

$\overline{V}_{mb_i}$ is the moving-bed velocity in the subsection i,

A_i is the cross-sectional area perpendicular to flow in subsection i, and

$Q_{measured}$ is the total unadjusted (biased) discharge for the transect.

5. Follow steps 2 through 4 for each transect used during the discharge measurement. The mean of the $Q_{corrected}$ for all transects is the corrected discharge for the measurement.

Average Moving-Bed Method

1. As with the subsection method, the first step is to compute the moving-bed velocity for each moving-bed test by dividing the DMG by time (eq. B6).

2. Sum the moving-bed velocities for each test and divide by two more than the number of moving-bed tests, which will account for the absence of a moving bed at both edges of water. Therefore, if three moving-bed tests were made, the sum of the moving-bed test velocities would be divided by 5.

3. The final step is to compute the corrected discharge. Calculate the discharge correction by multiplying the average moving-bed velocity computed in the previous step by the area of the cross section measured perpendicular to the mean flow direction. Then add the discharge correction to the measured discharge to determine the corrected discharge.

$$Q_{corrected} = Q_{measured} + (\overline{V}_{mb} A_{pf}), \qquad \text{(B8)}$$

where
$Q_{measured}$ is the total unadjusted (biased) discharge for the transect,
$\overline{V}_{mb}$ is the average moving-bed velocity in the measurement section (from step 2), and
A_{pf} is the average cross-sectional area perpendicular to flow in the transect.

Distributed Method using the SMBA Software

The distributed method uses the SMBA computer software to automate the correction process for StreamPro and Rio Grande ADCPs. The SMBA program can be obtained from the USGS at *http://hydroacoustics.usgs.gov/movingboat/mbd_software.shtml*. A distributed method similar to that used in the loop method (Mueller and Wagner, 2006) was developed for multiple moving-bed tests. Instead of using the mean moving-bed velocity measured from the loop method, the SMBA program distributes the average moving-bed velocity from the multiple stationary moving-bed tests to each ensemble based on the ratio of near-bed velocity for each ensemble and the mean near-bed velocity for the cross section. The distributed moving-bed velocities are then applied to the water and boat velocities for all depth cells in each of the corresponding ensembles in the measured portion of the cross section to determine the corrected measured discharge. The total discharge measured by an ADCP consists of discharge measured in a portion of the cross section and estimates of

discharge in the unmeasured top, bottom, left, and right edges. Therefore, the final corrected measured discharge is computed using the ratio of the corrected and uncorrected measured portion of the discharge to correct the sum of the measured and top and bottom estimated discharges. Water velocities near the bank are assumed to be sufficiently low as to not cause a moving bed; therefore, no correction is applied to the left and right edge discharges.

Mid-Section Method

The third GPS alternative is referred to as the mid-section method and uses the ADCP to measure discharge in a manner similar to a standard cup-meter measurement. The hydrographer collects 20–25 velocity profiles with the ADCP at selected locations across the stream. A stationary water-velocity profile is collected by holding the ADCP in a specific location for a specified time and then averaging the data to obtain a mean velocity profile or a depth-integrated mean velocity at that location. The mid-section method is not biased by moving-bed conditions because the ADCP is held stationary for each measurement and bottom tracking is not referenced; therefore, the velocity measured by the ADCP is only the water velocity. The boat reference must be set to "none" in the software when making a mid-section measurement in moving-bed conditions, or the measured velocity will still be biased by the moving bed.

SonTek/YSI and Teledyne RD Instruments have developed software that supports mid-section discharge measurements. The velocity and depth data are collected from the ADCP, and the discharge is computed by using the mid-section method. The width for each measurement section is computed as half the distance from the current section to the previous section, plus half the distance from the current section to the next section. This width is then multiplied by the depth measured during the velocity measurement to compute an area for each section. Discharge for each of the sections is then computed by multiplying the mean water velocity by the cross-sectional area, and then the incremental discharges are summed to determine the total discharge.

The advantages to using the mid-section method are (1) a GPS is not required, (2) correct velocities are measured, (3) discharge-measurements procedures are familiar to hydrographers, and (4) software is available to automate the procedure. The disadvantages are (1) positions must be measured manually, (2) angled flow can introduce errors in the velocity measurement and must be handled carefully, (3) appropriate sampling times are required for each measurement, (4) the ADCP must be kept stationary because ADCP movement is not corrected, and (5) the entire cross section is not measured.

Azimuth Method

The fourth GPS alternative is referred to as the azimuth method. The azimuth method is based on the fact that as an ADCP moves across the stream, a moving bed will cause the bottom-track-based ship track to be distorted in the upstream direction. If the azimuth between the starting and ending point of an ADCP transect is known, and the compass on the ADCP has been properly calibrated, the difference in azimuth of the course of the transect from that manually measured and that determined by the ADCP is a measure of the moving-bed velocity and can be used to compute a correction factor for the transect (fig. B-8). The advantages to using the azimuth method are that a GPS is not required and computations are simple. The disadvantages are (1) start and stop markers must be established, (2) the accuracy is dependent upon the measured azimuth and actual start and stop locations, (3) a compass calibration is required, (4) an accurate cross-sectional area projected perpendicular to the mean flow direction is required, (5) measured velocities are still biased low, (6) bottom tracking must be maintained, and (7) field applications have shown variable accuracy due to errors in azimuth measurements. *This method is difficult to accurately conduct in the field and is not recommended.*

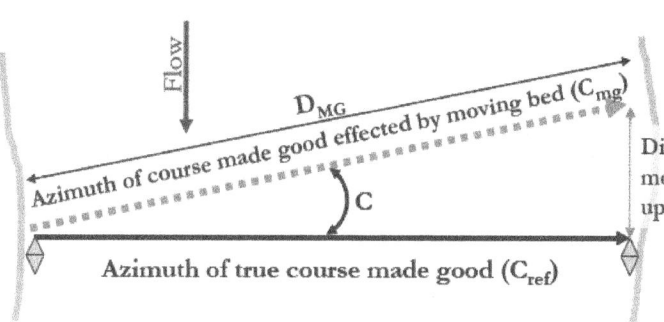

Figure B-8.   The azimuth method.

Field Procedures

1. Establish starting and ending points and mark with buoys or other markers. Start and stop the measurement as closely as possible to these points (within 1 foot, if possible).

2. Use a hand-held compass to measure the azimuth from one buoy or marker to the other, and record this number.

3. Calibrate the compass on the ADCP using the internal calibration routines.

4. Collect data as you would for a typical discharge measurement, but take special care to start and stop the transects at the buoys or markers.

Processing Procedures

1. Process the data as you would for a typical discharge measurement.

2. For each transect, note the DMG, the course made good (CMG), and the total time for measuring along the transect.

3. Determine the difference between the measured azimuth and the CMG, and name this difference, C. This angle should be in the upstream direction (fig. B-8).

$$C = \left| C_{mg} - C_{ref} \right|, \qquad \text{(B9)}$$

where

C is the angle between the manually measured and ADCP-measured course,

C_{mg} is the azimuth of CMG measured by the ADCP, and

C_{ref} is the azimuth of CMG measured manually with a compass.

4. The moving-bed correction factor for each transect can be computed as:

$$V_{mb} = \frac{D_{MG} \times \sin(C)}{T}, \qquad \text{(B10)}$$

where

V_{mb} is the moving-bed velocity,

D_{MG} is the DMG, and

T is the total measurement time for the transect.

5. Determine the cross-sectional area perpendicular to the mean flow direction (A) for the transect. The cross-sectional area will be affected by the moving bed unless corrected to be perpendicular to the mean flow direction.

6. Compute the corrected discharge for the transect

$$Q_{corrected} = Q_{measured} + (V_{mb}A), \qquad \text{(B11)}$$

where

V_{mb} is the moving-bed velocity for the transect,

A is the cross-sectional area perpendicular to flow for the transect, and

$Q_{measured}$ is the total unadjusted (biased) discharge for the transect.

7. Follow steps 2 through 6 for each transect used as part of the discharge measurement. The mean of the $Q_{corrected}$ for all transects is the corrected discharge for the measurement.

Appendix C – Description of Water Modes

ADCPs typically have a default water mode that can be used in a wide range of water conditions. Special configuration and processing of the acoustic pulse(s) reduce random errors in velocity measurements and allow velocity measurements to be made in shallow water but typically impose a restriction on the conditions that can be measured. Brief descriptions of the default and special water modes currently (2008) available for the SonTek/YSI RiverSurveyor and the TRDI Rio Grande are presented to provide users with a better understanding of how to optimize the use of an ADCP for making discharge measurements in a variety of site conditions.

SonTek/YSI RiverSurveyor Water Modes

The SonTek/YSI RiverSurveyor ADCP operates primarily with a single water mode, but uses a shallow-water ping to add data from near the water surface in low-velocity shallow-water conditions. The RiverSurveyor is a narrowband ADCP. Narrowband ADCPs are pulse-to-pulse incoherent ADCPs. This means that the ADCP transmits one simple pulse into the water, per beam per measurement (ping), and the resolution of the Doppler shift takes place during the duration of the received pulse. Velocity measurements made using the narrowband technology are noisy (have a relatively large random error). The RiverSurveyor compensates for the large random error by pinging fast (up to 20 Hz) and averaging many pings together before reporting a velocity. Typical response from a RiverSurveyor is a velocity-profile measurement every 5 seconds. Bottom-track pings are interleaved with the water pings once per second (SonTek/YSI, 2000).

The shallow-water ping is a pulse-coherent ping that is used in addition to the narrowband pings. The shallow-water ping measures the velocities in a depth cell that are above the standard velocity-profile depth cells. Because the shallow-water ping is a coherent ping, a potential exists for ambiguity errors. Ambiguity errors typically are avoided by the automatic parameters controlling the use of the ping, but the user can enable or disable the ping in the configuration part of the RiverSurveyor software. Enabling the shallow-water ping puts the ADCP in automatic mode where the shallow-water ping will only be used when there are three or fewer valid velocity-depth cells and the water velocity relative to the ADCP is less than 3.3 ft/s. The shallow-water ping is available on the 1.5-MHz and 3-MHz RiverSurveyors, which allow velocity data to be collected in water as shallow as 1.6 ft and 1 ft, respectively (SonTek/YSI, 2007).

TRDI Rio Grande and StreamPro Water Modes

The TRDI Rio Grande has five water modes, and the StreamPro has two water modes available to optimize the ADCP performance for the water velocity, turbulence level, and depth being measured. Because of the potential for large errors in the measured water velocity using Rio Grande water

mode 8, the use of water mode 8 is discouraged, and no discussion of this mode is provided. The other four available Rio Grande water modes include (1) water mode 1—general-purpose mode (2) water modes 5 and 11—low-velocity and low-turbulence modes, and (3) water mode 12—a fast ping-rate mode. Although the configuration software for the Rio Grande configures the appropriate mode for the user, the user should understand the operation and limitations of the modes in order to collect quality data, even in difficult conditions. The two water modes available in the StreamPro include a general-purpose mode similar to the Rio Grande water modes 1 and 12 and a low-velocity (less than 0.8 ft/s) mode.

Rio Grande Mode 1

Water mode 1 (WM1) is a general-purpose water mode for TRDI ADCPs (Teledyne RD Instruments, 2003). WM1 typically is used in streams with a mean depth deeper than 3.3 ft and(or) with velocities exceeding 3.3 ft/s. WM1 also can be used to measure slower velocities where water modes 5 and 11 will not work (see discussion of Rio Grande Modes 5/11). All other modes can be explained as a modification or enhancement of WM1. WM1 measures the Doppler shift using two phase-coded broadband pulses separated by a user-specified lag. The lag is inversely proportional to the radial ambiguity velocity, which is the maximum relative radial velocity (including boat speed and water speed) that can be accurately measured by the instrument. If the maximum boat velocity is assumed to equal the water velocity, an appropriate radial ambiguity velocity can be calculated to be approximately equal to the downstream water velocity (Teledyne RD Instruments, 2003). The recommended radial ambiguity velocity range is from 5.7 to 23 ft/s. The depth-cell size and lag between the pulses, and thus the ambiguity velocity, are key variables in determining the standard deviation of the random instrument noise present in velocity measurements. The recommended and commonly used depth-cell sizes for 600 kHz and 1,200 kHz instruments are 1.6 ft and 0.8 ft, respectively. This results in standard deviations of instrument noise of between 0.4 and 0.7 ft/s, depending on the radial ambiguity velocity value. The standard deviations will increase greatly for smaller depth-cell sizes.

Rio Grande Modes 5/11

Water modes 5 and 11 (WM5 and 11) are pulse-to-pulse coherent modes that use short phase-encoded broadband pulses. Similar to WM1, two pulses are transmitted; however, unlike WM1, the lag between the pulses for WM5 and 11 is long and variable. The lag is equal to the time for the first pulse to travel to the bottom and back. After the signal from the first pulse is received at the transducer face, the ADCP transmits the second pulse. The ADCP determines how long to wait before sending the second transmission from the water-depth-measurement part of the bottom-track measurement. This creates a very long lag with extremely low velocity single-ping standard deviations, typically less than 0.06 ft/s with depth-cell sizes of 0.16 ft and 0.33 ft for 1,200-kHz and 600-kHz instruments, respectively.

A long lag can cause a problem with residence time. Residence time is the time that a group of scatterers remains in a region for both pulses to ensonify the scatterers. If the velocity is very slow, most scatterers will remain in the same region for the time it takes both pulses to pass. Some decorrelation will occur because new scatterers enter the region as others leave. Nevertheless, if the number of scatterers entering and leaving is small, the correlation will be high and the data will be valid. If the velocity is too fast, and the scatterers move more than one-fourth to one-half of the transducer diameter with new scatterers introduced, the correlation between the two pulses will be low, and the data will be invalid.

The low-velocity standard deviations for WM5 and 11 make them excellent choices for discharge measurements where stream conditions permit use of these modes. The characteristics of these water modes that produce a low-velocity standard deviation also create significant limitations in the application of these water modes. Because of the long lag, the ambiguity velocity is very low and could render the modes nearly useless; however, an ambiguity-resolving depth cell is used to help resolve the ambiguity and allow a lower ambiguity velocity than the actual velocity of the water (Teledyne RD Instruments, 2003). The time-dilation technique used to determine the velocity in the ambiguity-resolving depth cell and the bin-to-bin tracking algorithm used to apply the ambiguity velocity to consecutive depth cells limits the use of WM5 and 11 to conditions with low turbulence and low shear. In WM5, the ambiguity-resolving depth cell extends from the end of the blank to the lesser of 2 ft or 85 percent of the shallowest beam, and the ambiguity-resolving depth cell must be at least 1 ft long. In WM11, the ambiguity-resolving depth cell is centered between the end of the blank and 85 percent of the shallowest beam and has a maximum length of 7.5 ft. Unlike WM5, which requires an ambiguity-resolving depth cell of at least 1 ft, WM11 continues to operate but stops computing ambiguity when the ambiguity-resolving depth cell becomes smaller than 1 ft. Shear caused by coarse bed material will often cause these modes to fail. Due to the short pulses and long lag, WM5 and 11 are limited to shallow depths (less than 13 ft for 1,200 kHz and less than 26 ft for 600 kHz) and slow velocities (typically less than 3.3 ft/s).

Rio Grande Mode 12

WM12 was designed to allow data collection in streams shallower than could be measured with WM1 and with velocities greater than could be measured with WM5 and 11. WM12 is essentially a fast ping rate WM1. The concept for WM12 is based on the fact that random noise is reduced by the square root of the number of samples. The velocity standard deviation increases as the WM1 depth-cell size is reduced. One method of reducing the velocity standard deviation is to collect and average more measurements. Averaging multiple WM1 pings (two pulses for each ping) only realizes gains in the transmission time of the data to the computer. WM12 is designed so that the heading and pitch-and-roll sensors

are only read at the beginning of the averaging period, the individual pings are averaged in phase space, and only the average is transformed into water velocities. This design eliminates some of the processing overhead and potential for averaging ambiguity velocity errors associated with WM1. The ping rate for WM1 is approximately 2–3 Hz, and the ping rate for WM12 is 10–20 Hz (depending on number of depth cells). However, since the heading and pitch-and-roll sensors are sampled only at the beginning of the averaging period, changes in heading, speed, pitch, or roll will lead to errors in the measured velocity. Thus, the sampling period needs to be short; generally 1 second or less is recommended. Although WM12 was designed for use with small depth cells in shallow water, WM12 can be used anywhere WM1 can be used, provided the ambiguity velocity is set properly as in WM1. The velocity standard deviation for WM12 cannot be stated as broadly as for the other water modes because WM12 is more configurable and the velocity standard deviation is dependent on the sampling period, the depth-cell size, the number of pings fit into the sampling period, and the ambiguity velocity.

StreamPro Mode 12

The StreamPro does not use water mode 12 as implemented in the Rio Grande, rather water mode 12 in the StreamPro (referred to hereafter as WM12sp) is more similar to a modified multi-ping Rio Grande water mode 1, which pings fast. For every water ping (WP) in WM12sp, the instrument sends and processes eight full WM1 pings. Each ping is fully processed, stored, and averaged with the other seven pings. Users cannot control the eight pings. The user can control the WP setting, which by default is set to 6 to obtain an ensemble output rate of 1 second. With default settings, 48 WM1 pings are averaged together and reported as the ensemble output. Ambiguity errors could occur in the individual pings and would be hidden by the averaging process; however, the StreamPro has a fixed ambiguity velocity of 11 ft/s, making an ambiguity error nearly impossible for the StreamPro applications. For WM12sp in the long-range mode, the number of pings processed is reduced to maintain an ensemble output rate of 1 second. The custom commands for WM12sp are limited to the blanking distance (0–1.6 ft), number of depth cells (0–20 depth cells) and depth-cell size (0.06–0.66 ft).

StreamPro Mode 13, Water Mode C, Low Noise Mode

The StreamPro has a second water mode that has a lower random noise than does WM12sp and that can be used to measure water velocities less than about 0.8 ft/s in water less than 3.3 ft deep. WM13 is a long-lag pulse coherent mode, which transmits multiple pulses at different lags. WM13 has no ambiguity-resolving depth cell as in Rio Grande WM5 and 11; the ambiguity in each depth cell is resolved using a proprietary approach.

Appendix D – Beam-Alignment Test

Introduction

One source of error in ADCP measurements is misalignment of beams in the instrument. This error can be checked and corrected by the user. The equations for both three-beam and four-beam (Appendix A) ADCPs assume that the beams are in perfect alignment and result in nominal transformation matrices for three-beam and four-beam systems. The nominal transformation matrix for a 25-degree three-beam system, such as the SonTek/YSI RiverSurveyor, is

$$\begin{bmatrix} 0 & -1.366 & 1.366 \\ 1.577 & -0.789 & -0.789 \\ 0.368 & 0.368 & 0.368 \end{bmatrix}.$$

The nominal transformation matrix for a 20-degree four-beam system, such as the TRDI Rio Grande, is

$$\begin{bmatrix} 1.4619 & -1.4619 & 0 & 0 \\ 0 & 0 & -1.4619 & 1.4619 \\ 0.2661 & 0.2661 & 0.2661 & 0.2661 \\ 1.0337 & 1.0337 & -1.0337 & -1.0337 \end{bmatrix}.$$

If the beams were misaligned during manufacturing, a custom transformation matrix to correct the misalignment is required. If the wrong transformation matrix is used, the water and bottom-track velocities will be consistently biased. The validity of the transformation matrix stored in the instrument can be determined by computing the ratio of the bottom-track and GPS straight-line distances over a long course, provided the instrument has a compass; for the StreamPro, no instrument rotation occurs during the test. Note: Procedures presented herein have not been validated in the field for StreamPro ADCPs.

Description of Procedure

The beam-alignment test is conducted by traversing a long (1,200–2,500 ft) course at a constant compass heading and speed while simultaneously recording GPS (GGA or VTG) and ADCP data. The length of the course depends on the accuracy of the GPS being used. The length of the course should be such that the error in GPS position is less than 0.1 percent of the length of the course. The ratio of the straight-line distance traveled (commonly called the DMG) as measured by bottom tracking with the ADCP and the straight-line distance traveled as measured by the GPS is computed. This ratio is referred to as the bottom-track-to-GPS ratio. A reciprocal traverse, which is a course of the same length at a heading approximately 180 degrees from the previous pass, is made and the ratios of the two passes are averaged. This procedure is repeated for a total of four times (eight passes altogether) while rotating the ADCP 45 degrees between each pair of courses. When the bottom-track-to-GPS ratio is less than 0.995, ADCP measurements most likely have a negative bias error, and when the bottom-track-to-GPS ratio is greater than 1.003, the ADCP most likely has a positive bias error (Oberg, 2002). A value for the bottom-track-to-GPS ratio of 0.995 corresponds to a –0.5-percent error in bottom-track velocity measurements. A value for the bottom-track-to-GPS ratio of 1.003 corresponds to a +0.3-percent error in bottom-track velocity measurements. The skewed criteria are due to a known potential for ADCPs to have a slight negative bias due to terrain effects. A well-calibrated ADCP should have bottom-track-to-GPS ratios of approximately 0.998 or 0.999. A form for documenting the beam-alignment tests is shown in figure D-1 (Oberg and others, 2005).

9/7/2005	U.S. DEPARTMENT OF THE INTERIOR *U.S. Geological Survey* **Acoustic Profiler Beam Alignment Test Form**	Processed by Ck'd by

Location					
Date	, 20____	Party			
Mfgr:	Model	Freq	Serial	Firmware	Software

ADCP Depth	Filename Prefix	Diagnostic Test File	Test Errors?
			Y or N

Boat/Motors Used	ADCP Water Temp	Measured Water Temp
	°C at	°C at

GPS Used	Moving Bed Test File	Moving bed
		Y or N

Site Conditions		Describe measurement location	
Max Water Depth			
Max Water Speed			
Max Boat Speed		Weather	
Water Mode			
Bottom Mode		Salinity	
Streambed material		ppt at	ppt at
Other Commands			

Beam Orientation	Heading	Filename or number	Field BT/GT	Final BT/GT	Notes
USGS Standard			0.995 to 1.003		
Comments					
			Sheet No	of	sheets

Figure D-1. Acoustic Doppler current profiler (ADCP) beam-alignment test form.

Step-by-Step Procedure

The following procedures should be followed when conducting the distance tests.

1. Conduct internal ADCP diagnostic tests (if available).

2. Lower the ADCP into the water, noting which beam is facing forward.

3. Using the data-collection software, begin pinging, but do not begin recording data.

4. Open a window in the software that will display the bottom-track-to-GPS DMG ratio.

5. Bring the boat to a constant speed and heading and note the heading. The speed should be fast enough to traverse the course in a reasonable time but not so fast as to cause invalid bottom-track data.

6. Once the boat is at the desired speed and heading, begin recording data. After traveling a minimum of 4,250 ft, record the bottom-track-to-GPS DMG ratio, stop recording, and then slow the boat and turn to a heading 180 degrees from the previous heading.

7. Bring the boat to a constant speed. Record data for this reciprocal pass. At the end of the pass, record the bottom-track-to-GPS ratio again. It is important to NOT slow the boat or change heading until recording is stopped.

8. Repeat this procedure while rotating the ADCP 45 degrees between each pair of courses until the ADCP has been rotated four times.

9. Average the bottom-track-to-GPS DMG ratio for each reciprocal pair.

10. Review the averaged bottom-track-to-GPS DMG ratio for all rotations and verify that all values are between 0.995 and 1.003. If values are outside of this range, have the instrument serviced by the manufacturer.

Appendix E – Forms and Quick-Reference Guides

Equipment Available	Equipment List
	Basic ADCP Equipment
	ADCP with attachments, bolts and nuts
	ADCP cable(s)
	Field computer with appropriate software
	Power cable/data port adapter for field computer
	Spare 12-V battery with appropriate wiring assembly
	Power inverters and power bars, if needed
	[illegible], or some other distance measurement device
	Battery charger
	ADCP measurement toolkit (see table [illegible])
	Field notes
	Tools
	Thermometer
	Multimeter
	Safety line
	Boat Deployment
	ADCP mount
	[illegible]
	Tethered / Remote Controlled (RCB) Boat Deployment
	Tethered boat and harness/float
	Long rope to use to tether the tethered boat
	Radio modems and cables
	Small 12-V, 9.6 batteries and charger
	Boat repair kit
	Bungee (for slow velocities)
	Weighted tether (for fast velocities)
	Hand-held walkie-talkie or phone radios
	DGPS Equipment
	DGPS and power/data cables
	DGPS antenna and cable
	Bracket for mounting DGPS antenna over ADCP
	12-V DC battery
	Spare fuses
	Echo Sounder
	Echo sounder and associated cables
	Mounting bracket for echo sounder
	Equipment for bar check
	12-V battery

Figure E-1. Pre-Field checklist of equipment for discharge measurements with acoustic Doppler current profilers (ADCPs).

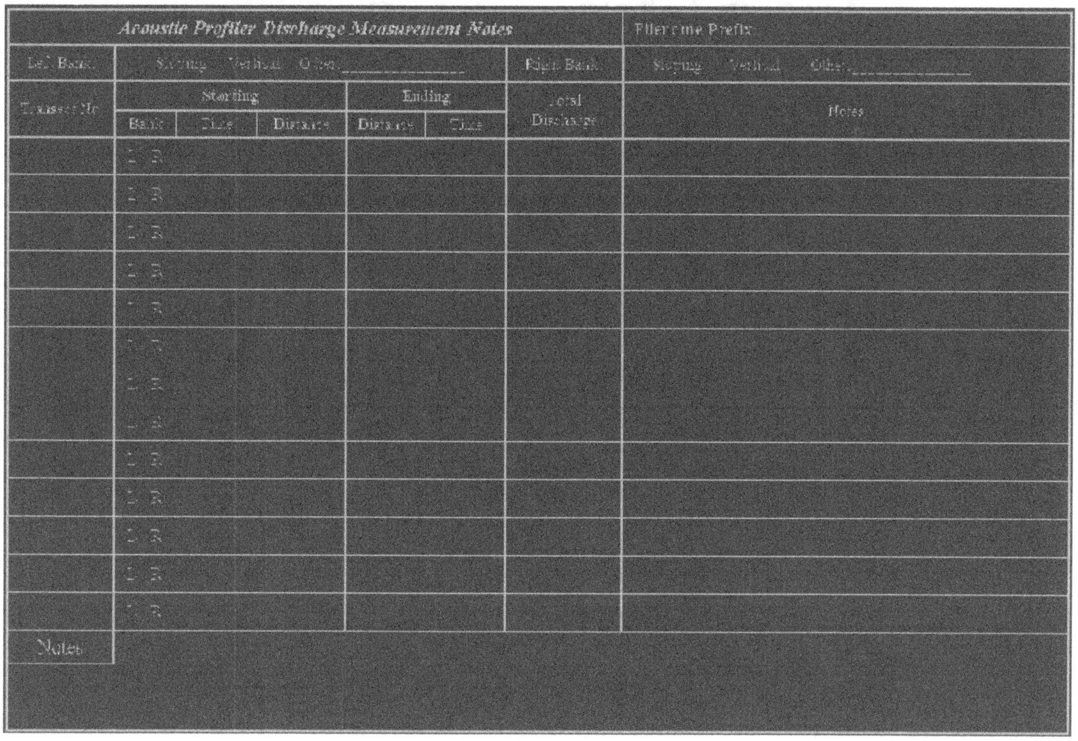

Acoustic Profiler Discharge Measurement Notes							Filename Prefix:
Left Bank:	Sloping Vertical Other______					Right Bank:	Sloping Vertical Other______
Transect Nr	Starting			Ending		Total Discharge	Notes
	Bank	Time	Distance	Distance	Time		
	L R						
	L R						
	L R						
	L R						
	L R						
	L R						
	L R						
	L R						
	L R						
	L R						
	L R						
	L R						
	L R						
Notes							

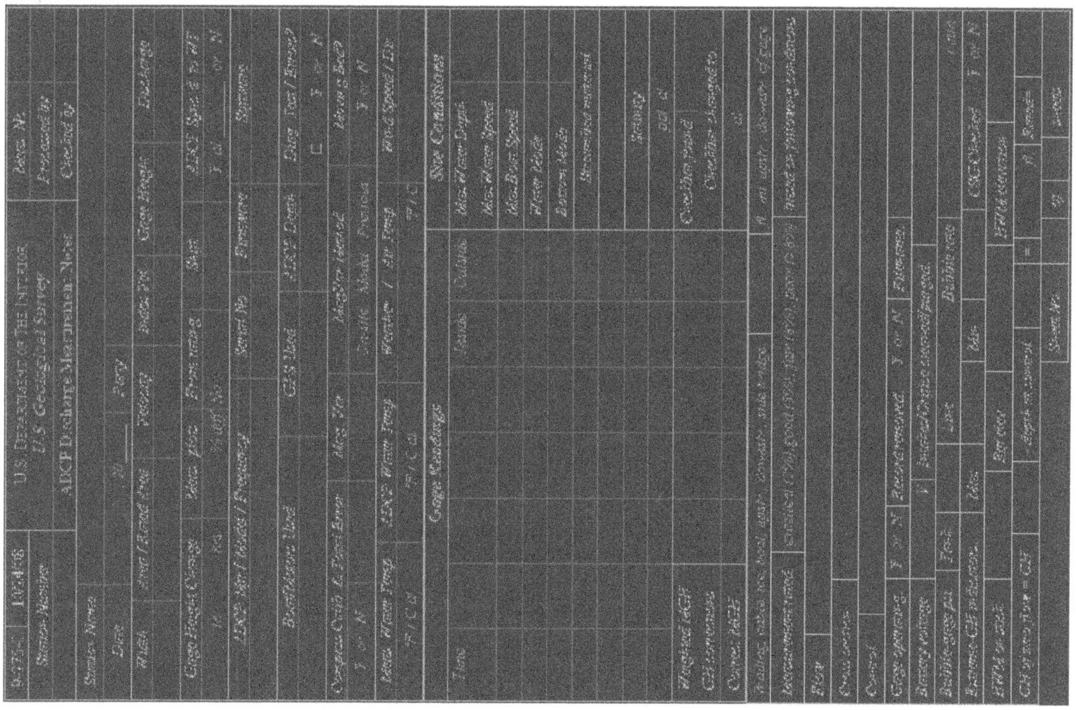

Figure E-2. Sample acoustic Doppler current profiler (ADCP) discharge-measurement field form.

Teledyne RD Instrument Rio Grande Quick Sheet

✓	DISCHARGE MEASUREMENT PROCEDURE
	1. Setup ADCP and Other Equipment
	a. Attach ADCP to mount or tethered boat
	b. Attach safety line to ADCP
	c. Turn on computer before connecting ADCP or data radios
	d. Turn off all automated field computer tasks/power saver settings
	e. Connect ADCP\GPS\field computer\data radios
	f. Verify communication with all devices
	g. Check and set ADCP clock time to appropriate time
	h. Measure water temperature, record, and compare to ADCP measured temperature
	2. Configure ADCP
	a. Locate appropriate measurement section / collect trial transect, if needed
	b. Select measurement site with uniform flow, no rapid drop-offs
	c. Minimize unmeasured area
	d. Determine maximum profiling depth
	e. Configure ADCP using automated software tools, if possible
	f. Measure salinity and if not zero, enter salinity in ADCP software
	g. Measure ADCP depth and record in software and notes (beware of pitch and roll)
	h. Fill out all field sheet with configuration and other information
	3. Prepare for discharge measurement
	a. Perform ADCP diagnostic tests and log results
	b. Perform and document compass calibration procedure (total error < 1 deg preferred)
	c. Record moving-bed test (stationary or loop)

	Stationary moving bed test	Loop test
	Duration of test = 600 seconds	Compass must be calibrated
	V_{mb}= Dist Upstream / Duration	Duration at least 3 minutes
	Moving bed if:	Boat speed less than 1.5 * water speed
	Anchored or tethered V_{mb}/V_w> 0.01	V_{mb}= Dist Upstream / Duration
	Not Anchored Boat V_{mb}/V_w> 0.02	Moving bed if:
	GPS Referenced V_{mb}/V_w> 0.01	V_{mb} > 0.04 ft/s and V_{mb}/V_w> 0.01
	V_w is the mean water velocity	V_w is the mean water velocity

✓	
	b. Use GPS or other appropriate technique, if a moving bed is present
	c. Establish start/stop points
	i. Need minimum of two depth cells with "good" velocity on each edge
	ii. May use buoys, pilings, poles, or other reference (avoid ferrous objects)
	3. Make discharge measurement
	a. Position boat at starting edge-of-water (two 'good' depth cells)
	i. Begin recording data
	ii. Measure and record distance to shore
	a. Hold position for minimum of 10 ensembles
	b. Drive boat across the river
	i. Boat speed should be less than or equal to the water speed
	ii. Be a smooth operator
	d. Approach ending shore slowly
	i. Hold position for minimum of 10 ensembles
	ii. Stop recording
	iii. Measure and record distance to shore
	iv. Collect 4 or more transects
	v. All transects must be within 5 percent of the mean discharge; except for unsteady flow conditions; if not, another set of transects should be measured and all transects collected averaged for the final discharge.
	e. Evaluate data in field, looking for potential problems in the data
	f. Make temporary backups before leaving the site

Figure E-3. Front page of quick reference guide for making acoustic Doppler current profiler (ADCP) discharge measurements with Teledyne RD Instruments Rio Grande profilers (modified from Oberg and others, 2005).

Recommendations and Limitations

Rio Grand Model >	1200/1200ZH	600
Blanking Distance (WF)		
All Modes	0.82 ft (25 cm)	0.82 ft (25 cm)
Minimum Depth Cell (Bin) Size		
Mode 1	0.82 ft (25 cm)	1.64 ft (50 cm)
Mode 5 or 11	0.16 ft (5 cm)	0.33 ft (10 cm)
Mode 12	0.16 ft (5 cm)	0.33 ft (10 cm)
Maximum Profiling Range		
Mode 1 or 12	65 ft	230 ft
Mode 5 or 11	13 ft	26 ft
Mode 5 or 11 with WZ3	22 ft	42 ft
Maximum Relative Velocity		
Mode 1 or 12	32 ft/s	32 ft/s
Mode 5 or 11	~2.3 ft/s	~3.3 ft/s
Mode 5 or 11 with WZ3	< 2.3 ft/s	< 2.3 ft/s
Approximate Velocity Standard Deviation		
Mode 1, WV175	*Bin Size: 0.82 ft* SD: 0.43 ft/s	*Bin Size: 1.64 ft* SD: 0.43 ft/s
Mode 5/11	*Bin Size: 0.16 ft* SD: < 0.03 ft/s	*Bin Size: 0.33 ft* SD: < 0.03 ft/s
Mode 12, WV175, 10 subpings	*Bin Size: 0.82 ft* SD: 0.13 ft/s	*Bin Size: 1.64 ft* SD: 0.13 ft/s
	Bin Size: 0.33 ft SD: 0.33 ft/s	*Bin Size: 0.82 ft* SD: 0.30 ft/s
	Bin Size: 0.16 ft SD: 0.16 ft/s	*Bin Size: 0.33 ft* SD: 0.49 ft/s

- **Modes 1, 5, and 11** should be configured for minimum bin size.
- **Mode 12** should be configured for a bin size smaller than the mode 1 bin size only as needed to measure in shallow areas. Mode 12 standard deviation will increase with a decreasing bin size.
- **Mode 12** should be configured to report data at least once per second and more often in turbulent conditions. This can be adjusted by setting the number of subpings in the WO command (WOss,tt, change the ss to a lower value). The standard deviation will increase as the number of subpings is lowered.
- **Modes 5 and 11** will not work in turbulent water or in sites with rough bottoms. If water is deep, the WZ03 can be used to extend the range, but then the turbulence and velocity must be very low. The precise velocity and turbulence limitations of modes 5 and 11 cannot be specified.

Missing Data

Lost Ensembles: Lost ensembles are a result of a communications problem.
Solution:
1. Disable antivirus, power management, etc.
2. Try lowering the baud rate.
3. Change serial ports or serial port adapters.

Bad Ensembles: Bad ensembles are a result of site conditions or water mode selection.
Solution:
1. Try a different cross section.
2. If using water modes 5 or 11, try water mode 12.
3. Use the bottom track tabular view to determine if bad ensembles are caused by bottom track. If so, try bottom mode 7 or a different cross section.

Baud Rates

ADCP Baud Rate: A baud rate lower than 38.4k baud will result in less data being collected. Set ADCP baud rate in BB-Talk using CB to set baud rate. 9600 – CB411, 19.2k – CB511, 38.4k – CB611, 57.6k – CB711, 115.2 – CB811. Set BB-Talk to "Send CK on Baud Rate change (CB command)".

GPS Baud Rate: The minimum acceptable GPS baud rate depends on the number of NMEA 0183 data types being output but the following are good general guidelines.

GPS Update Rate	Baud Rate
1 Hz	4800 or higher
5 Hz	19.2k or higher
10 Hz	38.4k or higher

Helpful Shortcuts

F4	Start/Stop Pinging
F5	Start/Stop Transect
F8	Toggle Bank
F9	Toggle Ensemble Header Tabular view
F11	Toggle Detailed Discharge/Composite Tabular view
F12	Toggle Discharge Summary Tabular view
Ctrl-B	Reference - Bottom Track
Ctrl-G	Reference - GPS (GGA)
Ctrl-V	Reference – GPS (VTG)
Ctrl-N	Reference - None

(when Measurement Control window selected)
Ctrl-K	Add Note

(when Ship Track plot selected)
Ctrl-PgDn	Scale Sticks Down
Ctrl-PgUp	Scale Sticks Up

Draft Measurement

(figure adapted from Environment Canada, 2004)

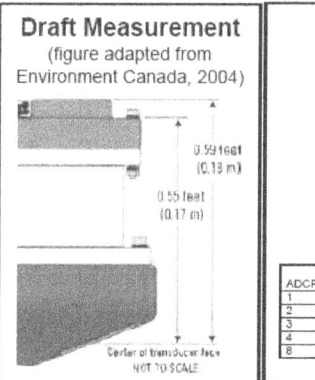

Cable Diagram

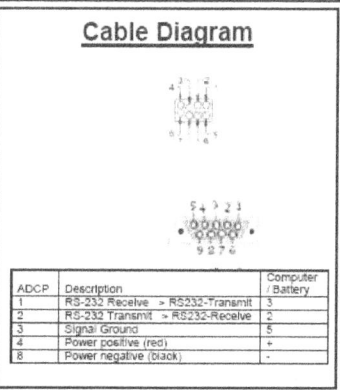

Figure E-4. Back page of quick reference guide for making acoustic Doppler current profiler (ADCP) discharge measurements with Teledyne RD Instruments Rio Grande profilers.

SonTek RiverSurveyor Quick Sheet

✓	DISCHARGE MEASUREMENT PROCEDURE
	1. **Setup ADCP and Other Equipment**
	a. Attach ADCP to mount or tethered boat
	b. Attach safety line to ADCP
	c. Turn on computer before connecting ADCP or data radios
	d. Turn off all automated field computer tasks/power saver settings
	e. Connect ADCP\GPS\field computer\data radios
	f. Verify communication with all devices
	g. Check and set ADCP clock time to appropriate time
	h. Measure water temperature, record, and compare to ADCP measured temperature
	2. **Configure ADCP**
	a. Locate appropriate measurement section / collect trial transect, if needed
	b. Select measurement site with uniform flow, no rapid drop-offs
	c. Minimize unmeasured area
	d. Determine maximum profiling depth
	e. Configure ADCP using automated software tools, if possible
	f. Measure salinity and if not zero, enter salinity in ADCP software
	g. Measure ADCP depth and record in software and notes (beware of pitch and roll)
	h. Fill out all field sheet with configuration and other information
	3. **Prepare for discharge measurement**
	a. Perform ADCP diagnostic tests and log results
	b. Perform and document compass calibration procedure (total error < 1 deg preferred)
	c. Record moving-bed test (stationary or loop)

<table>
<tr><td>Stationary moving bed test
Duration of test = 600 seconds
V_{mb}= Dist Upstream / Duration
Moving bed if:
  Anchored or tethered $V_{mb}/V_w > 0.01$
  Not Anchored Boat $V_{mb}/V_w > 0.02$
  GPS Referenced $V_{mb}/V_w > 0.01$
  V_w is the mean water velocity</td><td>Loop test
Compass must be calibrated
Duration at least 3 minutes
Boat speed less than 1.5 * water speed
V_{mb}= Dist Upstream / Duration
Moving bed if:
    $V_{mb} > 0.04$ ft/s and $V_{mb}/V_w > 0.01$
  V_w is the mean water velocity</td></tr>
</table>

✓	DISCHARGE MEASUREMENT PROCEDURE
	d. Use GPS or other appropriate technique, if a moving bed is present
	e. Establish start/stop points
	iii. Need minimum of two depth cells with "good" velocity on each edge
	iv. May use buoys, pilings, poles, or other reference (avoid ferrous objects)
	4. **Make discharge measurement**
	a. Position boat at starting edge-of-water (two 'good' depth cells)
	i. Begin recording data
	ii. Measure and record distance to shore
	b. Hold position for minimum of 10 ensembles
	c. Drive boat across the river
	i. Boat speed should be less than or equal to the water speed
	ii. Be a smooth operator
	d. Approach ending shore slowly
	i. Hold position for minimum of 10 ensembles
	ii. Stop recording
	iii. Measure and record distance to shore
	iv. Collect 4 or more transects
	v. All transects must be within 5 percent of the mean discharge; except for unsteady flow conditions; if not, another set of transects should be measured and all transects collected averaged for the final discharge.
	e. Evaluate data in field, looking for potential problems in the data
	f. Make temporary backups before leaving the site

Figure E-5.   Front page of quick reference guide for making acoustic Doppler profiler (ADP) discharge measurements with Sontek/YSI RiverSurveyor profilers (modified from Oberg and others, 2005).

Recommendations and Limitations

ADP Frequency (kHz)	Profiling Range [min. – max.] (ft)	Cell Size [min. – max.] (ft)	Blanking Distance [minimum] (ft)	Max. Bottom Tracking Depth (ft)
500	10 - 394	3.3 – 39.4	3.3	443
1,000	3.9 - 131	0.82 – 16.4	2.3	131
1,500	3.0 - 82	0.82 – 13.1	1.3	98
3,000	2.0 - 20	0.49 – 6.6	0.66	33

ADP Frequency (kHz)	Ping Rate (Hz)	Cell Size (ft)	Single Ping Std. Dev. (ft/s)	1-Second Std. Dev. (ft/s)	5-Second Std. Dev. (ft/s)
500	4.5	1.64	3.08	1.44	0.66
500	4.5	3.28	1.54	0.72	0.33
1,000	12	0.82	3.08	0.88	0.39
1,000	12	1.64	1.54	0.46	0.20
1,500	9	0.82	2.07	0.69	0.30
1,500	9	1.64	1.02	0.33	0.16
3,000	20	0.49	1.71	0.39	0.16
3,000	20	0.82	1.02	0.23	0.10

Draft Measurement
(adapted from Oberg and others, 2005)

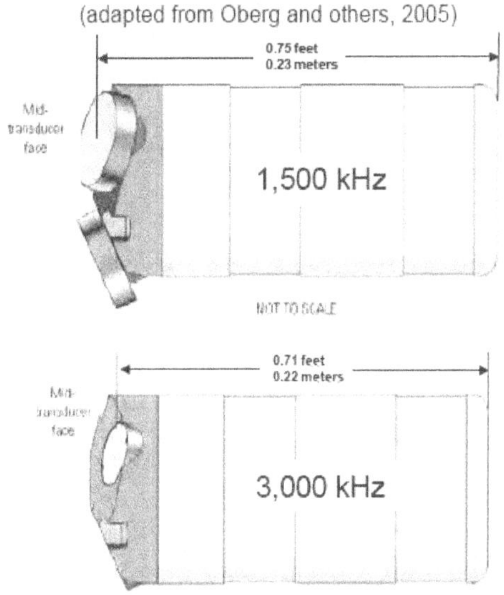

0.75 feet
0.23 meters

Mid-transducer face

1,500 kHz

NOT TO SCALE

0.71 feet
0.22 meters

Mid-transducer face

3,000 kHz

Helpful Shortcuts

F5	Start Pinging
F6	Stop Transect
F7	Start Recording
Alt-F7	Stop Recording
Ctrl-B	Reference - Bottom Track
Ctrl-G	Reference - GPS
Ctrl-E	English Units
Ctrl-M	Metric/SI Units
Ctrl-S	Communications Dialog
Ctrl-U	User Setup
Ctrl-H	Hardware Dialog
Ctrl-Y	Q Summary
Ctrl-D	Q Calculation Dialog
Ctrl-T	Q Report

+ (keypad) Scale Sticks Up
- (keypad) Scale Sticks Down

Stand-Alone ADP Connector Wiring
(adapted from Sontek, 2001)

IL-8-MP Pin No.	MIL-16-MP Pin No.	RS232	RS422
1	1	Vpower	Vpower
2	10	Data out	Tx+
3	11	Data in	Tx-
4	4 & 9	Drain	Drain
5	5	Not used	Not used
6	6	Not used	Rx+
7	14	Not Used	Rx-
8	16	Ground	Ground

Baud Rates

GPS Baud Rate: The minimum acceptable GPS baud rate depends on the number of NMEA 0183 data types being output but the following are good general guidelines.

GPS Update Rate	Baud Rate
1 Hz	4800 or higher
5 Hz	19.2k or higher
10 Hz	38.4k or higher

Figure E-6. Back page of quick reference guide for making acoustic Doppler profiler (ADP) discharge measurements with Sontek/YSI RiverSurveyor profilers.

Appendix F – Measurement Review Procedures

Teledyne RD Instruments Rio Grande and StreamPro ADCPs

1. **Load Data**
   - Load measurement file (*File→Open Measurement* or *Ctrl-O*).
   - Select Transect and *Reprocess Transect* (Shift+*F5*).

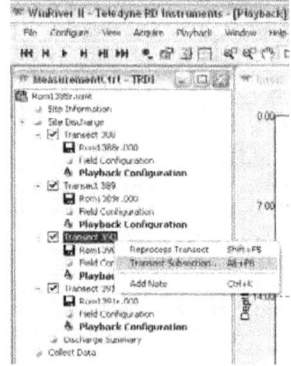

2. **Check Stick Ship Plots**
   - Look at stick ship plot.
   - Verify velocity reference (BT or GGA).
   - Step through depths using the down-arrow key (↓) (if problems are observed, look at velocity contour plots or intensity / back scatter plots).

3. **Velocity-Magnitude Contour**
   - Review velocity-magnitude contour.
   - Scale as appropriate.
   - Look for missing (referred to as "lost" in WinRiver) or invalid (referred to as "bad" in WinRiver) data, bed contour, lost ensembles, or other errors.

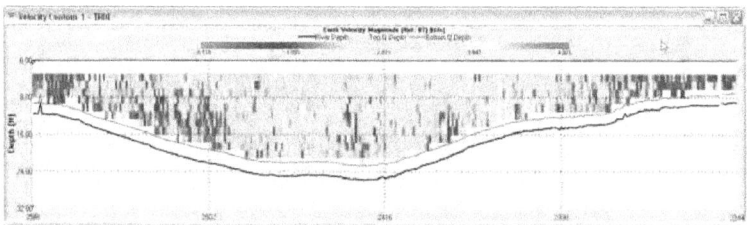

4. **Composite Tabular—Review**
   - Number of ensembles (total) vs
     - (1) Bad ensembles (invalid)
     - (2) Lost ensembles (missing).
   - What is value for "% Bad Bins (depth cells)?"
   - Is water temperature realistic?
   - Do the edge discharges appear reasonable?
   - Do edge discharges have correct signs?

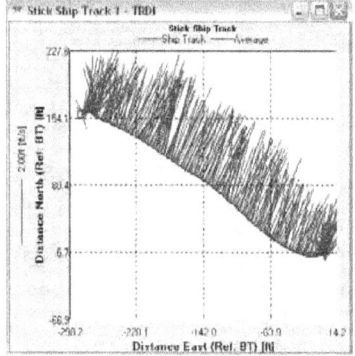

5. **System Parameters**
   - Press *F9* to check system parameters.
   - Does the information there (Frequency, Firmware, Water and Bottom modes, Bin Size) match the direct commands (*F3*) specified in the configuration file?
   - Does the information match the field form?

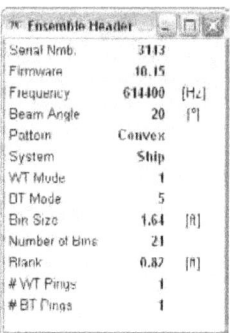

6. **Projected Velocity Contour**
   - Press *F2* to set angle for projected velocity contour plot and verify that value for the angle is correct. "Freeze" that value for the first transect—assuming that flow is uni-directional.
   - Scale as appropriate.
   - Look for reverse / bi-directional flow.

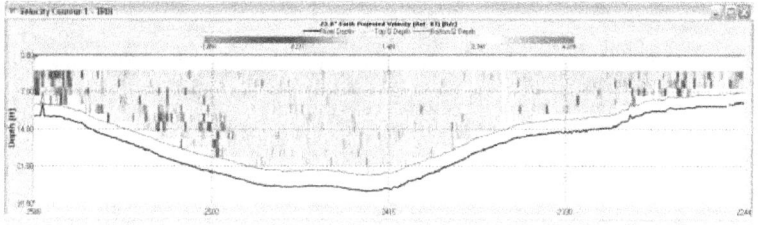

7. **Error Velocity**
   - Look for ambiguity errors, 3-beam solutions. Outliers can be related to ambiguity errors or turbulence.

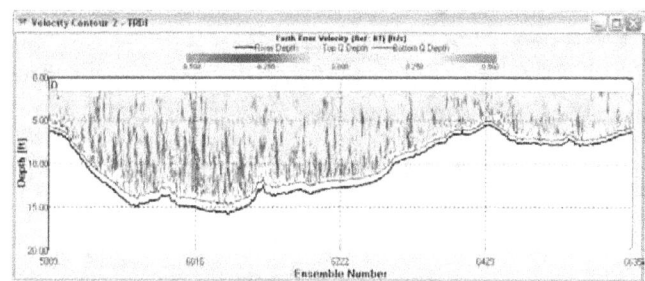

8. **Configuration – Direct Commands**
   - Press *F3* and check direct commands (Wizard and User).
   - Was the Config Wizard used?
   - Are the Wizard and User commands set correctly for ADCP and flow conditions?

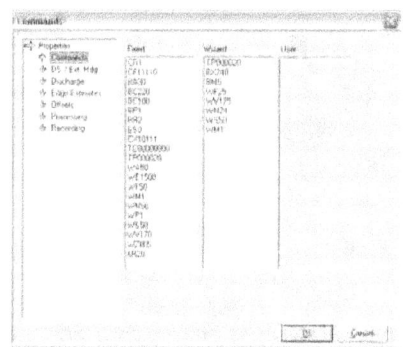

9. **Configuration – DS/GPS**
   - Check to see if *Depth Sounder* is used.

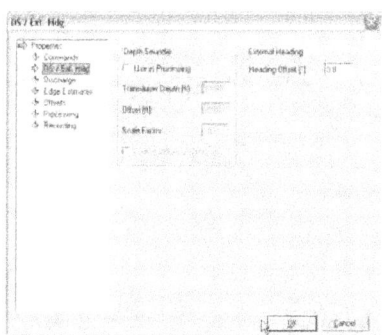

10. **Configuration – Discharge**
    - Are *extrapolation methods* correct?
    - Review *edge types* selected.
    - Were water profile depth cells cutoff? Why? Is it documented?
    - Correct number of *shore pings* (10)?

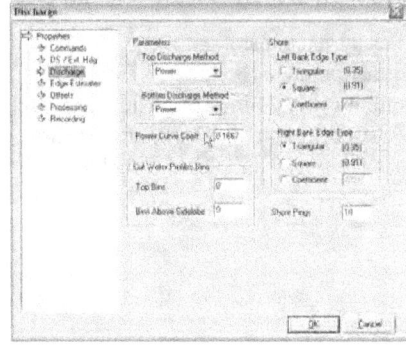

11. **Configuration – Edge Estimates**
    - Are *Edge Estimates* consistent with field form? If not, has explanation been supplied?
    - Are *Shore Distances* estimated rather than measured?
    - Are the estimated edge discharges reasonable for this section?

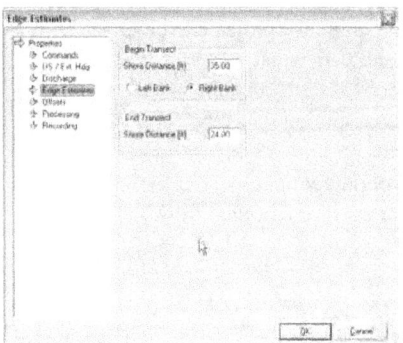

12. **Configuration – Offsets**
    - Does the *Transducer Depth* match the depth recorded on the field form?
    - Has *MagVar* been entered for discharge measurements made using GPS as a reference?

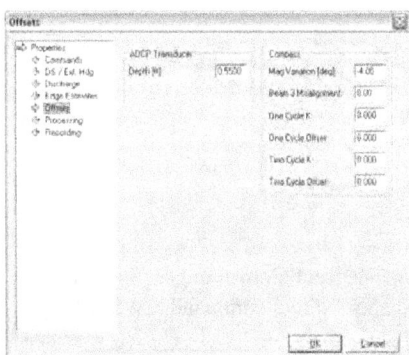

13. **Configuration – Processing**
    - Check area computation method.
    - Are thresholds used? If so, are they used correctly?
    - Is *Salinity* set correctly? Compare field form.
    - Are 3-beam solutions for the water- velocity (WT) data being used? (This is generally discouraged.)

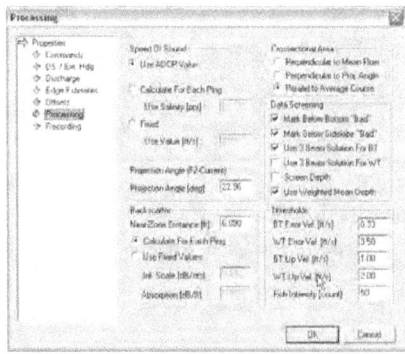

14. **Configuration – Recording**
   - GPS recorded?
   - Review comments field.

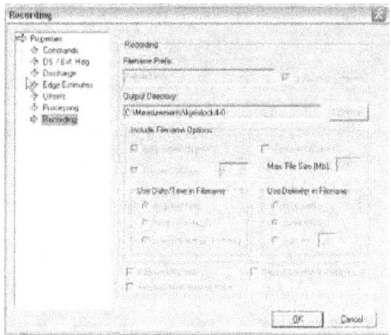

15. **Time Series**
   - Review time-series plots.
      o Compare *Water speed* and *Boat Speed* time series.
      o Review Pitch-and-roll plot for excessive pitch-and-roll.
   - Look for consistency, spikes, drop-outs, and large fluctuations.

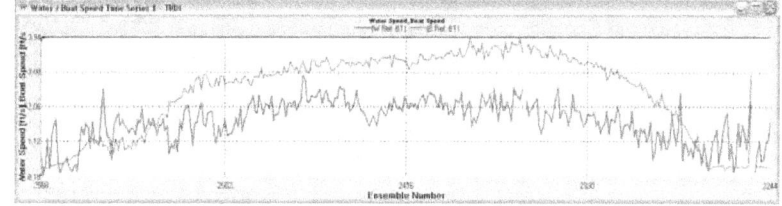

16. **Evaluate Extrapolation Method**
   - Open *View→Graphs→Profile Discharge* plot.
   - Average 10–20 ensembles (***Ctrl-F9***).
   - Modify extrapolation method as necessary.

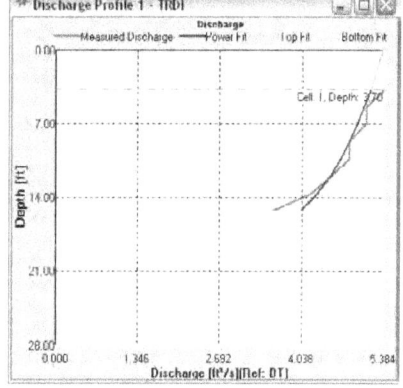

17. **Repeat**
   - Review all transects used during measurement.

18. **Check Whole Measurement**
   - Open *Discharge Summary* (***F12***).
   - Are all discharges within 5 percent? In other words, are any rows red?
   - Were reciprocal transect pairs obtained?
   - Check the following for consistency.
      i. Total area
      ii. Widths
      iii. Boat speed
      iv. Flow direction
      v. Duration
      vi. Compare boat speed with water speed

Example Discharge History Tabular for a "good" discharge measurement

Transect	Start Bank	# Ens.	Start Time	Total Q ft³/s	Delta Q %	Top Q ft³/s	Meas. Q ft³/s	Bottom Q ft³/s	Left Q ft³/s	Left Dist. ft	Right Q ft³/s	Right Dist. ft	Width ft	Total A ft²
MARS266	Right	891	10:41:47	5904.223	0.00	910.464	4500.925	490.255	13.949	25.00	32.631	29.00	571.92	5723.
MARS267	Left	914	10:49:39	5827.097	-1.52	910.447	4385.234	484.341	14.161	25.00	32.878	28.00	573.85	5689.
MARS268	Right	1206	10:59:10	6031.604	1.94	930.718	4502.479	540.032	20.094	25.00	38.246	28.00	571.63	5756.
MARS269	Left	735	11:10:34	5845.001	-1.22	900.030	4374.569	523.540	13.349	25.00	33.514	28.00	574.67	5767.
Average		936		5916.981	-0.00	914.915	4440.802	511.542	15.388	25.00	34.317	28.25	573.02	5734.
Std Dev.		196		97.691	1.65	12.960	70.459	24.981	3.156	0.00	2.645	0.50	1.48	34.8
Std. / Avg.		0.21		0.02	0.00	0.01	0.02	0.05	0.21	0.00	0.08	0.02	0.00	0.0

Example Discharge History Tabular for discharge measurement with "outliers"

Transect	Start Bank	# Ens.	Start Time	Total Q ft³/s	Delta Q %	Top Q ft³/s	Meas. Q ft³/s	Bottom Q ft³/s	Left Q ft³/s	Left Dist. ft	Right Q ft³/s	Right Dist. ft	Width ft	Total A ft²
MARS271	Left	806	11:35:06	5963.199	-7.36	903.208	4470.660	543.740	15.021	26.00	29.770	28.00	571.50	5764.
MARS272	Right	838	11:44:27	5887.979	-8.53	914.968	4395.299	526.577	20.588	25.00	30.618	28.00	573.47	5739.
MARS273	Left	737	11:51:38	6133.628	-4.71	922.560	4633.461	526.400	17.198	25.00	34.008	28.00	574.59	5865.
MARS274	Right	799	12:43:36	7762.128	20.59	1169.481	5881.163	641.985	28.322	25.00	41.177	28.00	571.31	5868.
Average		795		6436.734	0.00	977.554	4845.146	559.676	20.483	25.25	33.893	28.00	572.72	5809.
Std Dev.		42		889.552	13.82	128.198	697.793	55.472	5.597	0.50	5.189	0.00	1.59	67.2
Std. / Avg.		0.05		0.14	0.00	0.13	0.14	0.10	0.27	0.02	0.15	0.00	0.00	0.01

Sontek/YSI RiverSurveyor ADP

1. **Load Data**
   - Load measurement file (*File→Open Measurement* or *Ctrl-O*).
   - Select transect and *Reprocess Transect* (Shift+*F5*).

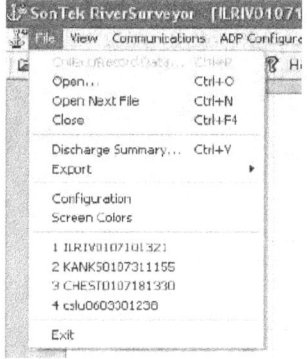

2. **Check Ship Plots**
   - Look at ship plot.
   - Verify velocity reference (BT or GPS).
   - Step through depths (if problems are observed, look at *velocity contour plots* or *intensity / SNR plots*).

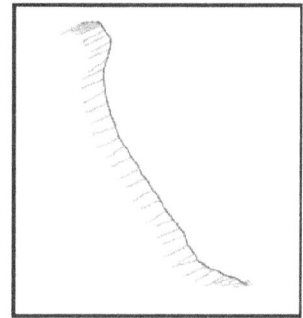

3. **Velocity-Magnitude Contour**
   - Review velocity-magnitude contour.
   - Scale as appropriate.
   - Look for invalid or missing data, bed contour, or other errors.

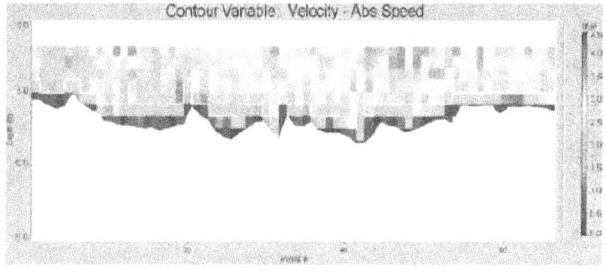

4. **Tabular Data**
    *   Number of profiles
    *   Is water temperature realistic?
    *   Do the edge discharges appear reasonable?
    *   Do edge discharges have correct signs?

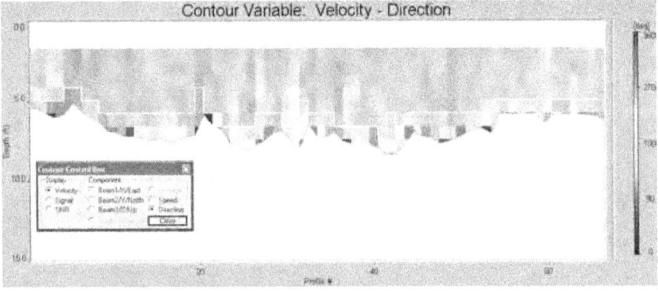

5. **Velocity-Direction Contour**
    *   Go to *View→Contour Control Box*. Select *Velocity* and *Direction* data.
    *   Scale as appropriate.
    *   Look for reverse / bi-directional flow.

6. **Configuration**
    *   Press ***Ctrl-U*** and check configuration.
    *   Check depth-cell size, range, and blank.
    *   Does the *Transducer Depth* match the depth recorded on the field form?
    *   Has *MagVar* been entered for discharge measurements made using GPS as a reference?

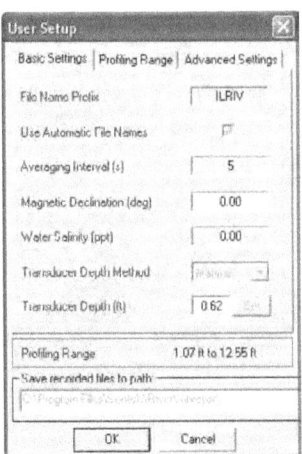

7. **Velocity Profile Extrapolation**
   - Select *Processing→Velocity Profile Extrapolation.*
   - Right-click on profile graph and select *profile extrapolation.*
   - Are *extrapolation methods* correct?
   - Review *edge types* selected.
   - Were water profile depth cells cut off? Why? Is it documented?

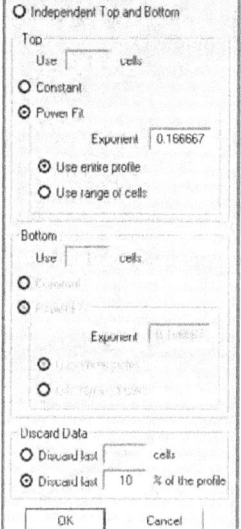
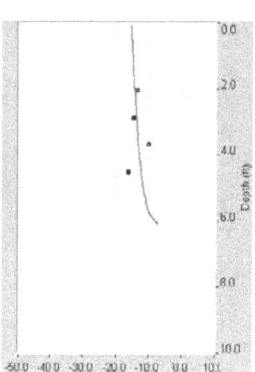

8. **Time Series**
   - Review time-series plots.
   - Right-click on the y-axis of the bar chart and select the appropriate variable.
   - Compare *Water Speed* and *Boat Speed* time series.
   - Review pitch-and-roll plot for excessive pitch-and-roll.
   - Look for consistency, spikes, drop-outs, and large fluctuations.

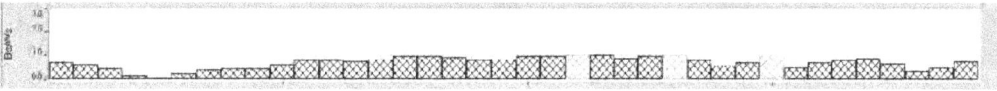

9. **Discharge Computation**
   - Select *Processing→Discharge Calculation.*
   - Are *Edge Distances* consistent with those recorded on the field form? If not, has explanation been supplied?
   - Are *Edge Distances* estimated rather than measured?
   - Are the estimated edge discharges reasonable for this section?
   - Recompute discharge.
   - Check for consistency and reasonableness.

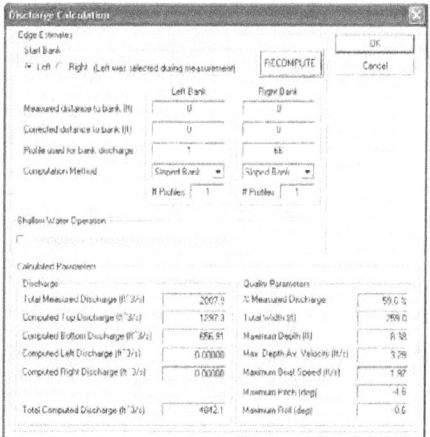

10. **Repeat**
    - Review all transects used for measurement.

11. **Check Whole Measurement**
    - Select *File→Discharge Summary* or ***Ctrl-Y***
    - Are all discharges within 5 percent?
    - Were reciprocal transect pairs obtained?
    - Check the following for consistency.
        - o Total area
        - o Widths
        - o Boat speed
        - o Flow direction
        - o Duration
        - o Compare boat speed to water speed